Marta Ruiz

Improving reordering and modeling in statistical machine translation

Marta Ruiz Costa-jussà

Improving reordering and modeling in statistical machine translation

Survey and new proposals

VDM Verlag Dr. Müller

Impressum/Imprint (nur für Deutschland/ only for Germany)

Bibliografische Information der Deutschen Nationalbibliothek: Die Deutsche Nationalbibliothek verzeichnet diese Publikation in der Deutschen Nationalbibliografie; detaillierte bibliografische Daten sind im Internet über http://dnb.d-nb.de abrufbar.

Alle in diesem Buch genannten Marken und Produktnamen unterliegen warenzeichen-, marken- oder patentrechtlichem Schutz bzw. sind Warenzeichen oder eingetragene Warenzeichen der jeweiligen Inhaber. Die Wiedergabe von Marken, Produktnamen, Gebrauchsnamen, Handelsnamen, Warenbezeichnungen u.s.w. in diesem Werk berechtigt auch ohne besondere Kennzeichnung nicht zu der Annahme, dass solche Namen im Sinne der Warenzeichen- und Markenschutzgesetzgebung als frei zu betrachten wären und daher von jedermann benutzt werden dürften.

Coverbild: www.purestockx.com

Verlag: VDM Verlag Dr. Müller Aktiengesellschaft & Co. KG
Dudweiler Landstr. 99, 66123 Saarbrücken, Deutschland
Telefon +49 681 9100-698, Telefax +49 681 9100-988, Email: info@vdm-verlag.de
Zugl.: Barcelona, Universitat Politècnica de Catalunya, 2008

Herstellung in Deutschland:
Schaltungsdienst Lange o.H.G., Berlin
Books on Demand GmbH, Norderstedt
Reha GmbH, Saarbrücken
Amazon Distribution GmbH, Leipzig
ISBN: 978-3-639-23568-5

Imprint (only for USA, GB)

Bibliographic information published by the Deutsche Nationalbibliothek: The Deutsche Nationalbibliothek lists this publication in the Deutsche Nationalbibliografie; detailed bibliographic data are available in the Internet at http://dnb.d-nb.de .

Any brand names and product names mentioned in this book are subject to trademark, brand or patent protection and are trademarks or registered trademarks of their respective holders. The use of brand names, product names, common names, trade names, product descriptions etc. even without a particular marking in this works is in no way to be construed to mean that such names may be regarded as unrestricted in respect of trademark and brand protection legislation and could thus be used by anyone.

Cover image: www.purestockx.com

Publisher:
VDM Verlag Dr. Müller Aktiengesellschaft & Co. KG
Dudweiler Landstr. 99, 66123 Saarbrücken, Germany
Phone +49 681 9100-698, Fax +49 681 9100-988, Email: info@vdm-publishing.com

Copyright © 2010 by the author and VDM Verlag Dr. Müller Aktiengesellschaft & Co. KG and licensors
All rights reserved. Saarbrücken 2010

Printed in the U.S.A.
Printed in the U.K. by (see last page)
ISBN: 978-3-639-23568-5

Agraïments

Aquesta tesi és el fruit de gairebé quatre anys de dedicació en els quals he tingut l'oportunitat d'aprendre i créixer tan a nivell personal com professional. Per això, ha estat una etapa molt intensa i enriquidora que engloba elements més enllà d'aquest document. Per tot plegat vull agrair el suport de les persones que m'han acompanyat al llarg d'aquests anys.

Moltes gràcies a l'Adrián per la seva guia indispensable. Són moltes les hores compartides, els dubtes resolts, les discussions i l'aprenentatge.

Així mateix, vull agrair al grup de traducció el seu acolliment. Especialment, al Pepe Mariño per la seva gestió del grup i la motivació que és capaç de transmetre. També a l'Adrià, al Josep Maria per atendre les meves consultes, al Patrik per la seva tranquil.litat i al Max.

Gràcies també als professors del departament i/o companys d'assignatura; als companys dels meus primers despatxos 216C i 218B que em van parlar de les seves experiències *doctorials*; als companys del 119 que essent el despatx d'estudiants més silenciós, tenim també el bon costum d'evadir-nos veient el *Cálico*; i al personal del departament. A tots ells gràcies per crear un bon ambient de treball.

No oblido pas les meves estades a París i les persones que van ser capaces d'influenciar-me positivament. La primera estada va ser decisiva en la meva elecció per començar el doctorat. La segona va enfocar el final de la tesi. Sincerament, espero poder tornar-hi una tercera vegada. À Jean-Luc Gauvain pour son accueil au LIMSI, à Olivier Galibert pour m'encourager à faire de la recherche, *vielen herzlichen Dank* Holger, et à tous mes collègues, ils étaient ma famille quand j'étais loin de chez moi: spécialement YY, Dani, María, Rena et Cristina.

Un element molt important que m'ha animat constantment han estat els amics amb els qui he pogut compartir aquest camí: el Jordi, la Mireia i el Jan, el Cristian, la Marta, el Marc, el Martí i l'Enric. És un privilegi estar envoltada de persones tan interessants i compromeses. Sens dubte, aquest és el començament d'una gran amistat.

Afortunadament, hi ha persones a les quals puc donar continuament les gràcies perquè quan et gires per demanar ajuda tens la certesa que sempre hi són: els meus pares i la meva germana Laura. També, al meu padrí Julià per la seva orientació. I a la família de l'Albert per interessar-se i il.lusionar-se per l'evolució d'aquesta tesi.

I a l'Albert, el meu marit, amb el qual compartim l'aventura de viure. Sincerament aquesta tesi és també teva pel teu constant recolzament, les teves crítiques constructives i les moltíssimes reflexions des de dalt de la *Mola*.

Contents

List of Figures

List of Tables

1

Introduction

This PhD thesis focuses on the framework of statistical machine translation (SMT), which is a specific approach to machine translation (MT). The main goal of MT is to be able to translate from a source language s to a target language t. MT is a difficult task, mainly because natural languages are highly complex. Many words have more than one meaning and sentences may have various readings. Certain grammatical relations in one language might not exist in another language. Moreover, there are non-linguistic factors such as the problem that performing a translation might require world knowledge. Additional challenges arise when dealing with spoken language translation like confronting non-grammatical texts.

In order to face the MT challenge, many dependencies have to be taken into account. Often, these dependencies are weak and vague, which makes it rarely possible to describe simple and relevant rules that hold without exception for different language pairs. SMT treats MT as a decision problem, where we have to decide upon several target sentences, given a source sentence and among all possible target sentences, we will choose the sentence with the highest probability according to a statistically-learned model. SMT technology has received increasing interest leading to improved algorithms and it has been justified by various successful comparative evaluations since its revival by the work of the famous IBM research group more than fifteen years ago. It has proved to be a competitive approach, which shows greater robustness than other methods for the translation of spontaneous speech. However, translations generated by SMT systems still have several significant challenges to pursue, like word reordering or word correspondences.

In order to understand the objectives and context of this work, this chapter delves into general aspects of SMT, including historical details, motivation and the challenges of the state-of-the-art SMT systems. Later, this chapter overviews the thesis content together with the project framework and main research contributions.

1.1   Motivation of Machine Translation and the Statistical Approach

At the end of the 19th century, L. L. Zamenhof proposed Esperanto; it was intended as a global language to be spoken and understood by everyone. The inventor was hoping that a common language could resolve global problems that lead to conflict. Esperanto as a planned language might have had some success, but today, the Information Society is and it will continue to be multilingual. For example, while surfing the Internet, one will sometimes come across languages and characters one does not understand. In this context, translation is the bottleneck of the pretended information globalization. MT may also be used in text or e-mail translation to the desired spoken language. Not only has MT created great expectations, but it also is the only solution to some situations. For example, Europe without MT is *lost in translation*. In Europe there are more than 20 official languages and this number continues to increase. The European Union institutions currently employ around 2,000 written-text translators and they also need 80 interpreters per language per day. Most translations regard to administrative reports, instruction manuals and other documents which have neither cultural nor literary value. In general, the demand for translations is increasing more quickly than the capacity of translators, apart from the problem of the lack of qualified candidates for some languages. As a consequence, machine translation is necessary.

The major approaches to machine translation are generally distinguished considering the level of linguistic analysis (and generation) required by the MT system. This can be graphically expressed by the machine translation pyramid in Figure 1.1. Typically, three different types of MT systems are identified: the *direct* approach, the *transfer* approach and the *interlingua* approach.

- The simplest approach, represented by the bottom of the pyramid, is the *direct* approach. Systems within this approach do not perform any kind of linguistic analysis of the source sentence in order to produce a target sentence.

- The *transfer* approach decomposes the translation process into three steps: analysis, transfer and generation. The source sentence is analyzed producing an abstract representation which is transferred into a corresponding representation in the target language. Finally, the generation step produces the target sentence from this intermediate representation. This approach was widely used in the 1980s, and despite great research effort, high-quality MT was only achieved for limited domains (Hutchins and Somers, 1992).

- The *interlingua* approach produces a deep syntactic and semantic analysis of the source sentence (language independent interlingual representation) in order to turn the translation task into the generation of a target sentence according to the obtained interlingual representation. This approach advocates for the deepest analysis of the source sentence. The interlingual language, once the source meaning is captured, can be expressed in any number of target languages, so long as a generation engine for each of them exists. The requirement that the whole source sentence needs to be understood before being translated, has proved to make the approach less robust to the ungrammatical expressions of informal language, typically produced by automatic speech recognition systems. However, it has been shown to work for limited domains.

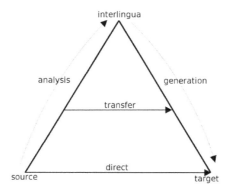

Figure 1.1: *Machine Translation pyramid.*

MT systems can also be classified according to their core technology. Under this classification we find *rule-based* and *corpus-based* approaches.

- In the *rule-based approach*, human experts specify a set of rules, to describe the translation process. This approach requires an enormous amount of input from human experts (Dorr, 1994; Arnold and Balkan, 1995).

- Under the *corpus-based approach*, the knowledge is automatically extracted by analyzing translation examples from a parallel corpus (built by human experts). The advantage is that, once the required techniques have been developed for a given language pair, (in theory) MT systems can be very quickly developed for new language pairs using provided training data.

  Within the corpus-based approaches we can further distinguish between *example-based* and *statistical* MT.

    - *Example-based* MT (EBMT) makes use of previously seen examples in parallel corpora. EBMT is often characterized by its use of a bilingual corpus with parallel texts as its main knowledge base, at run-time. It is essentially a translation by analogy and can be viewed as an implementation of case-based reasoning approach of machine learning.

    - In *Statistical* MT (SMT), parallel examples are used to train a statistical translation model. Thus, it relies on statistical parameters and a set of translation and language models, among other data-driven features. This approach initially worked on a word-by-word basis (hence classified as a direct method). However, current systems attempt to introduce a certain degree of linguistic analysis into the SMT approach, slightly climbing up the aforementioned MT pyramid.

Increasing computational power piqued the current interest in MT. Today the most widely used MT systems that try to provide automatic translations are: rule-based, example-based or statistical. The scientific community is really involved in MT and, particularly in SMT, which can be proved by the organization of major events such as evaluations (NIST [1],WMT [2], IWSLT [3], TC-STAR [4]) and the fact that outstanding research groups dedicate great efforts to its improvement. SMT feeds the computer with billions of words of text (bilingual aligned text

---

[1]http://www.nist.gov/speech/tests/mt/doc/mt06_evalplan.v4.pdf
[2]http://www.aclweb.org/anthology/W/W06/W06-15
[3]http://www.slc.atr.jp/IWSLT2006/
[4]http://www.tcstar.org/

consisting in human translations) and, then, SMT applies statistical learning techniques to build a translation model. The main advantages of SMT are cited as follows:

- SMT systems are not tailored to any specific pair of languages. To the translation system, any language is treated the same, and there is no manually created rule-set of grammar, metaphors and similar language considerations.

- There is a better use of resources. There is a great deal of natural language in machine-readable format. SMT relies on a large corpus of texts which are available in multiple languages.

- Generally, given that it is a corpus-based approach, SMT system is learning from existing human translations. Therefore, more natural translations have been achieved in international evaluations (as mentioned above) compared to other MT approaches.

1.2 Historical Overview of SMT: Past, Present and Future

Actually, statistical machine translation constitutes a research area inside the MT that has gained much attention worldwide during the last years. As of 2008, it is by far the most widely-studied machine translation paradigm. The first ideas of statistical machine translation were introduced by Weaver in 1949 (Shannon and Weaver, 1949), including the ideas of applying Shannon's information theory (Shannon, 1951) during World War II. According to this view, machine translation was conceived as the problem of finding a sentence by decoding a given "encrypted" version of it (Weaver, 1955). Although the idea seemed very feasible, enthusiasm faded shortly afterwards because of the computational limitations of the time (Hutchins, 1986). Finally, during the decade of the nineties, two factors made it possible for SMT to become realizable: first, a significant increase in both computational and storage capacity of computers, and second, the availability of large volumes of bilingual data.

Statistical machine translation was re-introduced in the early nineties by researchers at IBM's Thomas J. Watson Research Center where the first SMT systems were developed (Brown et al., 1990, 1993). These systems were based on the noisy channel approach, which models the probability of a target language sentence t given a source language sentence s as the product of a translation-model probability $p(s|t)$, which accounts for adequacy of translation contents, and a target-language probability $p(t)$, which accounts for fluency of target constructions. For these first SMT systems, translation-model probabilities at the sentence level were approximated from word-based translation models which were trained by using bilingual corpora.

In the case of target-language probabilities, these were generally trained from monolingual data by using n-grams. The first experiments were conducted on the proceedings of the Canadian Parliament, from English to French.

Present SMT systems have evolved from the original in primarily two ways: first, word-based translation models have been replaced by phrase-based translation models (Och and Ney, 2004; Zens et al., 2002; Koehn et al., 2003) which are directly estimated from aligned bilingual corpora by considering relative frequencies, and second, the noisy channel approach has been expanded to a more general maximum entropy approach in which a log-linear combination of multiple feature functions is implemented (Och and Ney, 2002).

As an extension of the machine translation problem, technological advances in the fields of automatic speech recognition (ASR) and text to speech synthesis (TTS) have made it possible to envision the challenge of spoken language translation (SLT) (Kay et al., 1992).

1.3 Thesis Content

This thesis began in Autumn 2004 in the Center of Speech and Language Applications and Technology (TALP) at the Universitat Politècnica de Catalunya (UPC). SMT had been set among the objectives in 2001 in the face of the promising horizon for Spoken Language Translation (SLT). Therefore, from 2001 to 2004 an initial translation engine based on Finite State Transducers had turned into an Ngram-based SMT system. However, the system still had several challenges to meet. Statistical machine translation must be able to handle compound words, idioms, morphology, syntax, out of vocabulary words and different word orders (the primary goal of this thesis). The last issue is one big difference between MT and Automatic Speech Recognition (ASR). While the speech signal and the corresponding textual representation can be mapped to each other in monotonic order, this is not always the case with the same text in two languages. For SMT, the translation model is only able to translate small sequences of words and word order must be taken into account because languages may differ in basic word order. Automatic translation between languages which do not share the same word-order type is confronted with this additional reordering challenge.

Thanks to team work, the UPC-TALP SMT system has achieved state-of-the-art results in several International Evaluation Campaigns which will be referred to in Appendix B.

1.3.1 Scientific Goals

This PhD is focused on achieving one main objective. Additionally, it addressed complementary challenges without deviating from the original scope.

1.3.1.1 Main Objective

- **To introduce novel reordering statistical techniques which are able to produce the translation in the correct word order.** Source and target languages may have different order structures. Languages may differ in the basic word order of verbs (V), subjects (S), and objects (O) in declarative clauses. For example, Spanish and English are both SVO languages. German, by contrast, is an SOV language while Classical Arabic and Urdu are VSO languages. Furthermore, there may be other structural differences in word orders between constituents; for instance, modifiers for nouns or verbs may be located in different places. Actually, an important deficit of current SMT systems is the difficult introduction of reordering capabilities. Incorporating them in the search process implies a high computational cost. However, reordering plays an important role, especially in some language pairs, such as Arabic or Chinese to English. The main issue of this thesis consists in developing an efficient reordering technique that can solve statistically the difference in word order of any language pair.

1.3.1.2 Complementary Objectives

- **To propose rescoring and system combination strategies which captures the best quality translations.** Current translation algorithms segment the given source sentence into units and then translate each unit. Therefore, it can become extremely complex to introduce feature functions that deal with information of the entire translated sentence. In those cases, translation may be performed in two steps. In the first step, we compute an N-best list. In the second step, the N-best list is reranked by applying additional features functions. Furthermore, the reranking of hypotheses allows for easier system combination. Different SMT systems approaches lead to different translations. In order to merge N-best lists which have been provided by different systems, we can use feature functions which decide which is the best translation. We contemplate developing and/or introducing feature functions to discriminate translations.

- **To gain efficiency and accuracy in the translation unit vocabulary.** By complementing the reordering objective, the extraction process and reordering techniques must be combined, either at the word or unit levels. Hence, the way the SMT system learns bilingual units plays an important role in translation quality. Here, the main idea is to further study the extraction of bilingual units from parallel corpora taking practical aspects, such as the translation vocabulary sparseness, into account. For some applications with limited memory space (PDAs, mobiles), the number of bilingual units stored in the device should be limited without affecting the quality of translation. In addition, the fewer bilingual units allows for faster translation time.

- **To build and maintain a state-of-the-art phrase-based system to compare its performance with the Ngram-based system.** We improved the SMT translation by using the Ngram-based system. To have impact and be relevant to the community, our improvements were demonstrated, when possible, in an in-house phrase-based system that is the most widely used system in SMT.

1.3.2   Organization

This PhD dissertation is divided into eight chapters. Following this introductory chapter, the second chapter describes the scientific framework. Chapters from 3 to 7 are dedicated to the description of contributions made by this research following a general chronological order. Finally, the last chapter concludes this work. Further details of each chapter are described below.

Chapter 2 presents a brief overview of the SMT approaches. Then, it details a study of the phrase- and Ngram-based SMT systems (§2.3.1 and §2.3.2) which have been used as core systems to develop this PhD work.

Chapter 3 brings together the research work done on enhancing the SMT systems presented in chapter § 2 by using continuous space language modeling (§3.2). This modeling applied to an SMT system takes further advantage of the limited data and adequately smooths word sequence probabilities.

Chapter 4 brings together the research work done on combining the SMT systems presented in chapter § 2. First, it reports an overview of the related work in system combination. Second, it describes a phrase- and Ngram-based system combination.

Chapter 5 reports a detailed description of the state-of-the-art algorithms used to meet the reordering challenge.

Chapter 6 presents the first reordering contribution, which was used in a language pair which required local reorderings. This approach is able to capture some local reorderings that current bilingual units cannot handle.

Chapter 7 describes the main contribution of this work: the SMR reordering approach. This approach is designed to capture local as well as global reorderings. SMR internal structure details are explored, e.g. the linguistic preprocessing (§7.2) and the coupling between the SMR and SMT systems (§7.3) varies. Additionally, this chapter provides comparison experiments (§7.4) against a very competitive reordering approach.

Chapter 8 draws the main conclusions from this PhD dissertation.

At the end of the document there are three appendices. **Appendix A** gives details of the corpora used throughout this work. **Appendix B** reports participation in International Evaluation Campaigns. **Appendix C** references the research contributions published throughout this PhD.

1.4   Project Framework

The research presented in this thesis is mainly based on work carried out in several research projects on spoken machine translation. Goals, run-time, funding and organization differ for each of the projects. However, machine translation is common to each of them. The following list includes brief descriptions of the projects involved:

Aliado (2004-2006)

Funded by the Spanish Government (TIC2002-04447-C02), the ALIADO project undertook the study and development of spoken and written language technologies for the design of personal assistants with a mobile terminal in a multilingual environment. The trade-off between the desire for a more powerful and a smaller terminal may be solved by means of an oral interface. Thus, attention was paid to the design of this interface and the problems related with environmental noise and the communication channel.

A broadcasting speech recognition system suitable to be integrated with the translator, a flexible dialog controller and a high quality text-to-speech converter with reduced memory requirements were all project goals. Furthermore, methodologies were developed for question understanding, and retrieval and summarization of information, so that the questions could be answered. Stochastic corpus-based approaches were considered in order to cope with the translation problem.

TC-STAR (2004-2007)

Funded by the European Union (IST-2002-FP6-506738), the TC-STAR project, financed by European Commission within the Sixth Program, was envisaged as a long-term effort to advance research in all core technologies for Speech-to-Speech Translation (SST). SST technology is a combination of Automatic Speech Recognition (ASR), Spoken Language Translation (SLT) and Text to Speech (TTS) (speech synthesis). The objectives of the project were ambitious: making a breakthrough in SST that significantly reduces the gap between human and machine translation performance.

The project targets a selection of unconstrained conversational speech domains (speeches and broadcast news) and three languages: European English, European Spanish, and Mandarin Chinese. Accurate translation of unrestricted speech was well beyond the capability of the state-of-the-art research systems. Therefore, advances were needed to improve these technologies for speech recognition and speech translation.

To foster significant advances in all SST technologies, periodic competitive evaluations were done. A measure of success of the project was the involvement of external participants in the evaluation campaigns. Results were presented and discussed in a series of TC-STAR evaluation workshops.

Avivavoz (2007-2009)

Funded by the Spanish Government (TEC2006-13964-C03), the Avivavoz project was proposed to perform advanced research in all key technologies related to speech translation systems (speech recognition, machine translation and speech synthesis).

The goal of the project was to achieve actual improvements in all speech translation system components in order to provide a speech mediating system for human communications among the official languages of the Spanish State (Spanish, Catalan, Basque and Galician), and between Spanish and English.

The project considered both improving and integrating all three technologies. In speech recognition, a robust system for a wide application domain (broadcast news and parliamentary sessions) and large vocabulary were developed. In machine translation, improvements were achieved for statistical translation techniques by including different sources of linguistic knowledge (event detection, syntactic and semantic analysis). In speech synthesis, new acoustic and prosodic models for expressive speech generation were developed. The final issue to be considered in this project was related to the interaction and integration of the three involved technologies.

All technologies developed in the project participated in competitive International Evaluation Campaigns, and the scientific and technological advances were shown by implementing a demonstrator system for automatic simultaneous interpretation of broadcast news and parliamentary speeches.

Tecnoparla (2007-2009)

Funded by the Catalan Government, the TECNOPARLA project is devoted to the multilingual society and its main goal is to further develop all key technologies related to speech translation systems (speech recognition, machine translation and speech synthesis). In this respect, it is similar to AVIVAVOZ project. The technology system will be built in the Spanish-Catalan task and the English-Catalan task. The last pair is challenging in the translation area because of limited available parallel data.

1.5   Research Contributions

Actively pursuing the major thesis objective has lead to the following main research contribution to the SMT field:

- A novel approach for solving the word reordering challenge in an SMT system: a first-pass translation is performed on the source-text, converting it to an intermediate representation, in which source-language words are presented in an order that more closely matches that of the target language. This first translation is performed using an Ngram-based system. Reordering is coupled with translation, which then allows a choice among multiple reordering paths.

While the main work is defined above, further research contributions are briefly described. The first is a preliminary reordering approach and the following ones are related to the additional thesis goals.

- A first limited novel approach for solving local word reorderings in an SMT system. The main limitation is that it addresses reordering in a deterministic way (a fixed reordering is given to the SMT system). However, good results are reported in the Spanish to English task.

- Experimental work to introduce continuous space language models both in phrase- and Ngram-based SMT systems and its influence in translation.

- A study of two state-of-the-art SMT systems mentioned above. This study leads to system combination at the rescoring level.

- Construction of several SMT systems which were presented at International Evaluation Campaigns. Building a machine translation system is a serious undertaking. The participants are usually provided with a common set of training and test data. Therefore, systems are evaluated under similar conditions, generally with automatic and human measures.

The research developed in this PhD has been published in a number of publications which will be referred to in their respective chapters and/or in Appendix §C.

2

Scientific Framework: SMT Baseline Systems

This chapter is an overview of the most significant issues in statistical machine translation systems that are relevant to our research. Section 2.1 outlines the mathematical fundamentals of SMT and introduces the maximum entropy approach, which is widely used to combine different sources of information. Section 2.2 introduces the concept of word alignment, which is a key issue in modeling string translation probability. Section 2.3 describes the current SMT translation models, which consider sequences of words as their translation units. We detail the phrase- and Ngram-based SMT systems. Emphasis is placed on the latter as it was developed in our research group. Then Section 2.4 describes several feature functions which, together with the translation model score, act as the translation units. Section 2.5 details how feature functions are combined and weighted using an optimization algorithm. To conclude this chapter, Section 2.6 reports the most common ways to evaluate a SMT system according to automatic and human metrics.

2.1 Introduction

The main goal of SMT is the translation of a text given in some source language into a target language. We are given a source string $s_1^J = s_1 \ldots s_j \ldots s_J$, which is to be translated into a target string $t_1^I = t_1 \ldots t_i \ldots t_I$. Among all possible target strings, we will choose the string with the highest probability:

$$\tilde{t}_1^I = \underset{t_1^I}{argmax} \, P(t_1^I | s_1^J) \tag{2.1}$$

The first SMT systems were reformulated using Bayes's rule. This approach, called the noisy channel, models the probability of a target language sentence $t_1^I = t_1 \ldots t_I$ given a source language sentence $s_1^J = s_1 \ldots s_J$ as follows:

$$\tilde{t}_1^I = \underset{t_1^I}{argmax} \, \frac{p(s_1^J | t_1^I) \, p(t_1^I)}{p(s_1^J)} \tag{2.2}$$

We can ignore the denominator $p(s_1^J)$ inside the *argmax* function since we are choosing the best t_1^I sentence for a fixed sentence s_1^J, and hence $p(s_1^J)$ is a constant.

$$\tilde{t}_1^I = \underset{t_1^I}{argmax} \, p(s_1^J | t_1^I) \, p(t_1^I) \tag{2.3}$$

Here $p(t_1^I)$ is the language model of the target language. A language model is usually formulated as a probability distribution over strings that attempts to reflect how likely a string occurs inside a language (Chen and Goodman, 1998). Statistical MT systems make use of the same n-gram language models as do speech recognition and other applications. The language model component is monolingual, so acquiring training data is relatively easy. Then, $p(s_1^J | t_1^I)$ is the string translation model, which is the basis of the translation. Two translation models will be described in Section 2.3.1 and 2.3.2. The *argmax* operation denotes the search problem, i.e. the generation of the output sentence in the target language.

In recent systems, such an approach has been expanded to a more general maximum entropy approach in which a log-linear combination of multiple feature functions is implemented (Och, 2003). With respect to criterion 2.3, this approach leads to maximising a linear combination of feature functions:

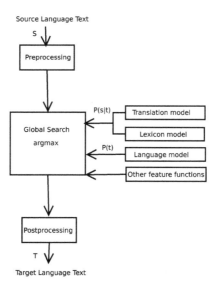

Figure 2.1: *Architecture of the translation approach based on the log-linear framework approximation.*

$$\tilde{t} = \underset{t}{argmax} \left\{ \sum_{m=1}^{M} \lambda_m h_m(t, s) \right\} \tag{2.4}$$

The overall architecture of this statistical translation approach is summarized in Figure 2.1. This framework leads to the two approaches considered in this work: the phrase- and the Ngram-based. The main difference between the two approaches is in how they deal with the translation model. These systems are detailed in the following sections.

2.2   Statistical Word Alignment

A key issue in modeling the string translation probability $p(s_1^J|t_1^I)$ is how we define the correspondence between words of the target and source sentences. In typical cases, we can assume a sort of pairwise dependence by considering all word pairs (s_j, t_j) for a given sentence pair (s_1^J, t_1^I). A word alignment is a mapping between the source words and the target words in a set of parallel sentences. Given parallel texts, the task of automatic word alignment focuses on detecting which tokens or set of tokens from each language are connected together in a given translation context.

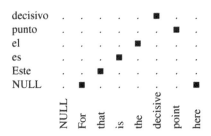

Figure 2.2: *Word alignment example.*

Figure 2.2 shows a visualization of an alignment between the English sentence *For that is the decisive point here* and the Spanish sentence *Este es el punto decisivo*. Additionally, this figure shows what is called the word alignment matrix.

The alignment approach that we used in this thesis is an instance of unsupervised learning, where the system is not given examples of the kind of output desired, but it is instead trying to find values for the alignments which best explain the observed bi-text. We only assume that we already know which sentence in the target text aligns with which sentence in the source text.

In principle, we can have arbitrary alignment relationships between the target and the source words. So called statistical alignment models are decomposed into:

- *fertility model.* It accounts for the probability that a target word t_i generates ϕ_i words in the source sentence.

- *lexicon model.* It models the probability to produce a source word s_j given a target word t_i.

- *distortion model.* It tries to explain the phenomenon of placing a source word in position j given that the target word is placed in position i in the target sentence (also used with inverted dependencies, and known as the alignment model).

The different combinations of these three models are commonly known in literature as IBM models (Brown et al., 1993). Currently, word alignments based on IBM and HMM models for which a systematic performance comparison can be found in (Och, 2003), are considered to be the state of the art. Typically, the implementation by (Och and Ney, 2000), which is freely-available in the GIZA++ package, is used. As mentioned, nearly all current approaches to statistical translation depend on the results of these alignment processes to estimate their translation models, such as (Och, 1999; Koehn et al., 2003; Mariño et al., 2006), among others.

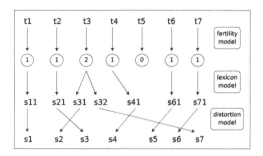

Figure 2.3: *Illustration of the generative process underlying IBM models.*

IBM models 1 and 2 do not include fertility parameters and their likelihood distributions are guaranteed to achieve a global maximum. Their difference is that Model 1 assigns a uniform distribution to alignment probabilities, whereas Model 2 introduces a zero-order dependency with the position in the source. (Vogel et al., 1996) presented a modification of Model 2 that introduced first-order dependencies in alignment probabilities, the so-called HMM alignment model, with successful results. Model 3 introduces fertility and Model 4 and 5 introduce more detailed dependencies in the distortion model. All of them must be numerically approximated and not even a local maximum can be guaranteed.

A detailed description of IBM models and their estimation from a parallel corpus can be found in (Brown et al., 1993). An informal yet clarifying tutorial on IBM models can be found in (Knight, 1999b).

As explicitly introduced by the IBM formulation as a hidden random variable, word alignment becomes a function from source positions j to target positions i, so that $a(j) = i$. This definition implies that the resultant alignment solutions will never contain many-to-many links, but only many-to-one[1], as only one target position is possible for a given source position j.

Although this limitation does not account for many real-life alignment relationships, in prin-

---

[1]By many-to-many links those relationships between more than one word in each language are referred, whereas many-to-one links associate more than one source word with a single target word. One-to-one links are defined analogously.

ciple IBM models can solve this by estimating the probability of generating the source empty word, which can translate into non-empty target words.

In 1999, the John Hopkins University summer workshop research team on SMT released GIZA (as part of the EGYPT toolkit), a tool implementing IBM models training from parallel corpora and best-alignment Viterbi search, as reported in (Al-Onaizan et al., 1999), where a decoder for model 3 is also described. This was a breakthrough that enabled many other teams to join SMT research easily. In 2001 and 2003, improved versions of this tool were released, and were named GIZA++(Och and Ney, 2003).

However, many current SMT systems do not use IBM model parameters in their training schemes, but instead use only the *most probable* alignment (using a Viterbi search) given the estimated IBM models (typically by means of GIZA++). Therefore, in order to obtain many-to-many word alignments, usually alignments from source-to-target and target-to-source are performed, applying symmetrization strategies. Several symmetrization algorithms have been proposed, the most widely known being the **union, intersection** and **refined** (Och and Ney, 2000) of source-to-target and target-to-source alignments, and the **grow-diag-final**(Koehn et al., 2005), which employs the previous intersection and union alignments.

Lately, recent work has begun to explore supervised methods which rely on presenting the system with a (usually small) number of manually aligned sentences (Callison-Burch et al., 2004). Besides the benefit of the additional information provided by supervision, these models are able to combine many features of the data, such as context, syntactic structure, part-of-speech, or translation lexicon information, which are difficult to integrate into the generative statistical models traditionally used (Fraser and Marcu, 2006; Lambert, 2008). Additionally, discriminative alignment training allows to extend the IBM models with new (sub)models, which leads to additional increases in word-alignment accuracy.

In order to evaluate the quality of the word alignment task, the Alignment Error Rate (AER) measure as proposed in (Och and Ney, 2000) has been commonly used. Given a manual gold standard alignment with the criterion of Sure and Possible links, we can define Recall, Precision and AER measures thus:

$$recall = \frac{|A \cap S|}{|S|}, \ precision = \frac{|A \cap P|}{|A|}$$

$$AER = 1 - \frac{|A \cap S| + |A \cap P|}{|A| + |S|} \tag{2.5}$$

where A is the hypothesis alignment and S is the set of Sure links in the gold standard reference, and P includes the set of Possible and Sure links in the gold standard reference.

In (Fraser and Marcu, 2006), the authors confirm experimentally that the AER does not correlate well with MT performance. On the other hand, they obtain strong correlations with the F-Measure:

$$\text{F-measure with Sure and Possible}(A, P, S, \alpha) = \frac{1}{\frac{\alpha}{Precision(A,P)} + \frac{(1-\alpha)}{Recall(A,S)}} \tag{2.6}$$

The basic property of F-Measure is that unbalanced precision and recall should be penalized.

2.3 Current SMT Translation Models

The job of the translation model, given a target sentence and a foreign sentence, is to assign a probability that t_1^I generates s_1^J. While we can estimate these probabilities by thinking about how each individual word is translated, modern statistical MT is based on the intuition that a better way to compute these probabilities is by considering the behavior of phrases (sequences of words). The intuition of phrase-based statistical MT is to use phrases as well as single words as the fundamental units of translation. Phrases are estimated from multiple segmentation of the aligned bilingual corpora by using relative frequencies. The main difference between the phrase- and word-based models is that the former manages bilingual units of several words (i.e. discurso extenso # long speech) instead of only individual words themselves.

The translation problem has also been approached from the finite-state perspective as the most natural way for integrating speech recognition and machine translation into a speech-to-speech translation system (Vidal, 1997; Bangalore and Riccardi, 2000b; Casacuberta, 2001). The Ngram-based system implements a translation model based on this finite-state perspective (de Gispert and Mariño, 2002) which is used along with a log-linear combination of additional feature functions (Mariño et al., 2006).

In the next lines we introduce these two state-of-the-art translations models, which work at the phrase level. Afterwards we explain the feature functions that are used together with the translation model in the log-linear framework.

2.3.1   Phrase-based Model

The basic idea of phrase-based translation is to segment the given source sentence into units (hereafter called phrases), then translate each phrase and finally compose the target sentence from these phrase translations.

Basically, a bilingual phrase is a pair of m source words and n target words. For extraction from a bilingual word aligned training corpus, two additional constraints are considered:

1. the words are consecutive, and,

2. they are consistent with the word alignment matrix.

This consistency means that the m source words are aligned only to the n target words and vice versa. The phrase-based approach was first presented in (Och, 1999) and named *Alignment Templates*, consisting of pairs of generalized phrases which allow for word classes and include internal word alignments.

A simplification of this model is the so-called phrase-based statistical machine translation presented in (Zens et al., 2002). This approach does not use word classes but instead uses bilingual phrases without internal alignment. The following criterion defines the set of bilingual phrases BP of the sentence pair $(s_1^J; t_1^I)$ that is consistent with the word alignment matrix A.

$$BP(s_1^J, t_1^I, A) = \{(s_j^{j+m}, t_i^{i+n}) : \forall (i', j') \epsilon A : j \leq j' leq j + m \leftrightarrow i \leq i' \leq i + n\} \qquad (2.7)$$

Figure 2.4 shows an example of a word-aligned sentence pair and the bilingual phrases extracted from this sentence pair according to the defined criteria.

The extraction of bilingual phrases from a word alignment corpus can be done in a straightforward manner and pseudo-code is reported in (Zens et al., 2002). To use the bilingual phrases in the translation model, the hidden variable B is introduced. This is a segmentation of the sentence pair $(s_1^J; t_1^I)$ into K phrases $(s_K^J; t_1^K)$. We use a one-to-one phrase alignment, i.e., one source phrase is translated by exactly one target phrase.

$$Pr(s_1^J | t_1^I) = \alpha(t_1^I) \cdot \sum_B Pr(\tilde{s}_k \mid \tilde{t}_k) \qquad (2.8)$$

where the hidden variable B is the segmentation of the sentence pair in K bilingual phrases $(\tilde{f}_1^K, \tilde{e}_1^K)$, and $\alpha(t_1^I)$ assumes equal probability for all segmentations.

Given the collected phrase pairs, the phrase translation probability distribution is commonly estimated by relative frequency in both directions.

$$P(s|t) = \frac{N(s,t)}{N(t)} \qquad (2.9)$$

$$P(t|s) = \frac{N(s,t)}{N(s)} \qquad (2.10)$$

where N(s,t) means the number of times the phrase s is translated by t.

PHRASE-BASED MODELING ISSUES

- **Observation pruning.** This pruning can be defined as any technique taking a decision on discarding certain training material because they are almost useless. The phrase translation model becomes huge when increasing training material. A typical pruning is limiting the maximum phrase length. How long do phrases have to be to achieve high performance? (Koehn et al., 2003) reports results where limiting the length to a surprising level already achieves top performance. Learning longer phrases does not yield much improvement, and occasionally leads to worse results. Reducing the limit too much, however, is clearly detrimental. Allowing for longer phrases increases almost linearly the phrase translation table size.

  Additionally, the phrase vocabulary may be pruned by keeping the N most frequent phrases with common source sides.

  Phrase pruning has a direct incidence on the overall system performance. While a low value of N will significantly decrease translation quality, on the other hand, a large value of N will provide the same translation quality than a more adequate N would, but with a significant increment in computational costs. The optimal value for this parameter depends on data and should be adjusted empirically for each considered translation task.

2.3.2   Ngram-based Model

Another approach using sequences of words instead of single words in the translation model is the so-called Ngram-based approach. Again, it regards translation as a stochastic process and the search of the translation maximizes the joint probability $p(f,e)$. Differently to the phrase-based translation model, the Ngram-based translation model is trained on bilingual n-grams. This model constitutes a language model of a particular *"bi-language"* composed of bilingual units (translation units) which are referred to as **tuples**. In this way, the translation model probabilities at the sentence level are approximated by using n-grams of tuples, such as described by the following equation:

$$\hat{t}_1^I = \arg \max_{t_1^I}\{p(s_1^J, t_1^I)\} = \cdots = \tag{2.11}$$

$$\arg \max_{t_1^I}\{\prod_{n=1}^{N} p((s,t)_n|(s,t)_{n-x+1}, \cdots, (s,t)_{n-1})\} \tag{2.12}$$

where the n-*th* tuple of a sentence pair is referred as $(s,t)_n$.

As any standard n-gram language model, the bilingual translation model is estimated over a training corpus composed of sentences in the language being modeled. In this case, we consider sentences in the *"bi-language"* previously introduced.

The Ngram-based approach is monotonic in that its model is based on the sequential order of tuples during training. Therefore, the baseline system may be more appropriate for pairs of languages with relatively similar word order schemes.

TUPLE SEGMENTATION

Tuples are extracted from a word-to-word aligned corpus in such a manner that a unique segmentation of the bilingual corpus is achieved. Although in principle, any Viterbi alignment should allow for tuple extraction, the resulting tuple vocabulary strongly depends on the particular alignment set considered. According to our experience Mariño et al. (2006) in some specific tasks, the best performance is achieved when the union of the source-to-target and target-to-source alignment sets (IBM models (Brown et al., 1993)) is used for tuple extraction.

In this way, different from other implementations, where one-to-one (Bangalore and Riccardi, 2000a) or one-to-many (Casacuberta and Vidal, 2004) alignments are used, tuples are extracted from many-to-many alignments. This implementation produces a monotonic segmentation of bilingual sentence pairs, which allows for the simultaneous capture of contextual and the reordering of information into the bilingual translation unit structures. This segmentation also allows for the estimation of the n-gram probabilities appearing in

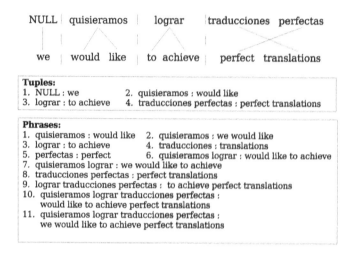

Figure 2.4: *Example of tuple and phrase extraction.*

(2.12). In order to guarantee a unique segmentation of the corpus, tuple extraction is performed according to the following constraints:

- a monotonic segmentation of each bilingual sentence pair is produced,

- no word inside the tuple is aligned to words outside the tuple, and

- no smaller tuples can be extracted without violating the previous constraints.

Notice that according to this, tuples can be formally defined as the set of shortest phrases that provides a monotonic segmentation of the bilingual corpus. Figure 2.4 presents a simple example illustrating the unique tuple segmentation for a given pair of sentences, as well as the complete phrase set.

Two important observations from Figure 2.4 must be considered:

1. **NULL.** The first important observation from Figure 2.4, is related to the possible occurrence of tuples containing unaligned elements in its target side. This is the case of

tuple 1. This kind of tuples should be handled in an alternative manner for the system to be able to provide appropriate translations for such unaligned elements. The problem of how to handle this kind of situation is discussed in detail in (Mariño et al., 2006). In short, since no NULL is actually expected to occur in translation inputs, this type of tuple is not allowed. Any target word that is linked to NULL is attached either to the word that precedes or to the word that follows it. To determine this, we use the IBM-1 probabilities. More specifically, the IBM-1 lexical parameters (Brown et al., 1993) are used for computing the translation probabilities of two possible new tuples: the one that results when the null-aligned-word is attached to the previous word, and the one that results when it is attached to the following one. Then, the attachment direction is selected according to the tuple with the highest translation probability.

2. **Embedded words.** The second observation from Figure 2.4, is that it often occurs that a large number of single-word translation probabilities are left out of the model. This happens for all words that are always embedded in tuples containing two or more words. Consider for example the word "translations" in Figure 2.4. This word is embedded into tuple 4. If a similar situation is encountered for all occurrences of "translations" in the training corpus, then no translation probability for an independent occurrence of this word will exist.

To overcome this problem, the tuple n-gram model is enhanced by incorporating 1-gram translation probabilities for all the embedded words detected during the tuple extraction step. These 1-gram translation probabilities are computed from the intersection of both the source-to-target and the target-to-source alignments.

Currently, much research efforts is being directed towards adapting the Ngram-based approach to be used for language pairs with different word orders as will be explained in Chapter § 5.

A complementary approach to translation with reordering can be followed if we allow for a certain reordering in the training data. This means that the translation units are modified so that they are not forced to sequentially produce the source and target sentences anymore. The reordering procedure in training tends to monotonize the word-to-word alignment through changing the word order of the source sentences.

The rationale of this approach (hereafter, called unfolding) is double, on the one hand, it makes sense when applied into a decoder with reordering capabilities. On the other hand, the unfolding technique generates shorter tuples, alleviating the problem of embedded units (tuples only appearing within long distance alignments, not having any translation in isolation). The unfolding technique (Crego et al., 2005a) uses the word-to-word alignments as follows:

- First, an iterative procedure, where words on one side are grouped when linked to the same word (or group) on the other side. The procedure loops grouping words on both sides until no new groups are obtained.

- Second the resulting groups (unfolded tuples) are extracted, keeping the word order of target sentence words, even though, the tuples sequence modifies the source sentence word order. See an example in Figure 2.5.

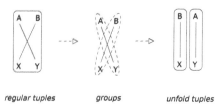

regular tuples   groups   unfold tuples

Figure 2.5: *Different bilingual tuples are extracted using the regular (default) and unfolding methods.*

NGRAM-BASED MODELING ISSUES

Here, we focus on the *n*-gram implementation of the tuples bilingual model, conducting a brief reference of its main modeling aspects (de Gispert, 2006).

- **History length.** Regarding model size, increasing the history length has a direct correlation with storage and computational costs. History length is associated with the size of the data structures needed by an *N*-gram-based decoder in order to produce translations. Depending on the task, trigrams or 4-grams seem to be the best option.

- **Observation pruning.** Another important observation from Figure 2.4 is that each tuple length is implicitly defined by the word-links in the alignment. Different from phrase extraction procedures, for which a maximum phrase length should be defined in order to avoid a vocabulary explosion, tuple extraction procedures do not have any control over tuple lengths. According to this, the tuple approach will greatly benefit from the structural similarity between the languages under consideration. Then, for close language pairs, tuples are expected to successfully handle those short reordering patterns which are included in the tuple structure, as in the case of "perfect translations : traducciones perfectas" presented in Figure 2.4. On the other hand, in the case of distant pairs of languages, for which a large number of long tuples are expected to occur, this baseline approach will more easily fail to provide a good translation model due to tuple sparseness.

Similar to the phrase case, in our Ngram-based SMT system implementation, the tuple vocabulary is pruned by using histogram counts. This pruning is performed by keeping the N most frequent tuples with common source sides.

- **Smoothing the bilingual model.** Smoothing refers to all techniques that redistribute probability from seen events to unseen events. Different smoothing methods follow different strategies for removing probability mass from seen events (discounting) and assigning it to unseen combinations (back-off) (Jelinek, 1998). Due to the bilingual nature of the translation model, a comparative experiment was conducted in (de Gispert, 2006) in order to assess which smoothing was best suited for the bilingual N-gram modeling. The modified Kneser-Ney presented in (Chen and Goodman, 1998) achieves the best translation scores. It leads to the conclusion that the bilingual model behaves similarly to a standard (monolingual) language model when it comes to smoothing techniques.

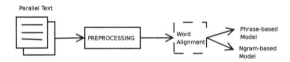

Figure 2.6: *Training steps of the translation model.*

Figure 2.6 draws how the translation models (either phrase or the Ngram-based) are trained.

2.4   Feature Functions

Currently, most SMT systems consider a log-linear framework of probabilistic information (Och, 2002; Mariño et al., 2006; Matusov et al., 2006b; Bertoldi et al., 2006).

Following Figure 2.1, the straightforward approach in a phrase-based system would be the following combination of feature functions (in this case, it is similar to the noisy channel):

$$h_{tm}(t_1^I, s_1^J) = log\ p(s_1^J|t_1^I) \quad h_{lm}(t_1^I, s_1^J) = log\ p(t_1^I)$$ (2.13)

Here, $p(t_1^I)$ denotes the trained language model and $p(s_1^J|t_1^I)$ denotes the trained translation model. We obtain two maximum entropy model parameters λ_{lm} and λ_{tm} that can be trained using an optimization algorithm.

Following Equation 2.13, the straightforward approach to defining the feature functions for the maximum entropy model in a Ngram-based system would be the definition of the following feature function:

$$h_{tm}(t_1^I, s_1^J) = log\ p(s_1^J, t_1^I)$$ (2.14)

Here, the translation model already includes a language model and no additional features are necessary. However, the maximum entropy model can consider qualitative differences of the different component models when introducing more refined feature functions as follows:

- A word bonus for each produced target word.

$$h_{wb}(t_1^I, s_1^J) = I$$ (2.15)

- Additional language models introducing linguistic knowledge as POS target language model (Crego et al., 2006c; Costa-jussà et al., 2006). Also we could include language models built with neural networks, see §3.2.

- A distortion model. Distortion in statistical machine translation refers to a word having a different ('distorted') position in the target sentence than the corresponding word had in the source sentence. The feature functions related to distortion or reordering are reviewed and proposed in the next chapter.

- Lexical models (such as IBM model 1 from source to target and from target to source).

$$h_{ibm_1}(t_1^I, s_1^J) = log \frac{1}{(I+1)^J} \prod_{j=1}^{J} \sum_{i=0}^{I} p(t_j^n|s_i^n)$$ (2.16)

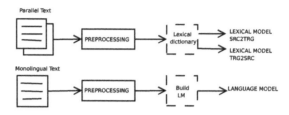

Figure 2.7: *Training steps of the most common feature functions.*

Figure 2.7 draws how to train these feature functions (except for the word bonus). As an additional language model, we use a target and POS target language model (when POS is available). This feature, together with the word penalty and the IBM-1 lexical model, is a straightforward integration into a dynamic programming search algorithm. When this is not feasible, as in the case of the language model using neural networks, we may use *n*-best rescoring for other features as is explained in Chapter 3.

2.5   Minimum Error Training

To train the model parameters λ_1^M of the direct translation model according to Eq. 2.4, a Minimum Error Rate iterative strategy is followed (Och, 2003).

An adequate algorithm for such a task is the *downhill simplex* algorithm (Nelder and Mead, 1965) or the SPSA (Lambert and Banchs, 2006). Our optimization tool was based on the downhill simplex algorithm. The method uses a geometrical figure called a simplex consisting, in N dimensions, of $N + 1$ points and all their interconnecting line segments, polygonal faces, etc. The starting point is a set of $N + 1$ points in parameter space, defining an initial simplex. At each step, the simplex performs geometrical operations (reflexions, contractions and expansions) until a local minimum is reached. In our case, it adjusts the log-linear weights so as to maximize an objective function. Note that in this problem, only a local optimum is usually found. Tuning is performed according to MT quality measures and evaluated over development data.

Two optimization schemes are possible. In the first one (depicted in Figure 2.8), the development corpus is translated at each iteration. With four parameters (one parameter can remain fixed to 1, the others being scaled accordingly), the algorithm converges after about 50 or more iterations. Thus, in this scheme, on the order of 50 development corpus translations are required.

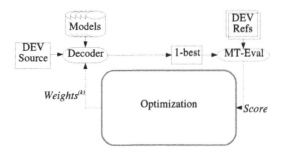

Figure 2.8: *Single-loop optimization block diagram.*

In the second scheme (depicted in Figure 2.9), an n-best list is produce by the decoder. The optimization algorithm is used to minimize the translation error while rescoring this n-best list. With the optimal coefficients, a new decoding is performed so as to produced an updated n-best list (Bertoldi, 2006). This process converges after only 5 to 10 decodings. For each internal optimization, about 50 iterations are still required, but each iteration is much shorter since they only require to rescoring an n-best list.

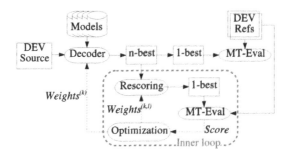

Figure 2.9: *Double-loop optimization block diagram.*

2.6   Evaluation Measures

Evaluating the quality of a translation is an extremely subjective task. Nevertheless, evaluation is essential, and research on evaluation methodology has been important from the earliest days of MT to the present. This section explains the different evaluation metrics (both automatic and manual) used for assessment of translation quality in this thesis.

2.6.1   Automatic Evaluation

The utility and attractiveness of automatic metrics for MT evaluation has been widely recognized by the MT community. Evaluating an MT system using such automatic metrics is much faster, easier and cheaper compared to human evaluations, which require trained bilingual evaluators. In addition to their utility for comparing the performance of different systems on a common translation task, automatic metrics can be applied on a frequent and ongoing basis during system development in order to guide the development of the system based on concrete performance improvements.

An automatic metric has to be both effective and useful. The most intuitive requirement is that the metric has very high correlation with quantified human notions of MT quality. Furthermore, a good metric should be as sensitive as possible to differences in MT quality between different systems, and between different versions of the same system.

Most efforts have focused on strategies for computing some kind of similarity score between the output of an MT system and one or more reference translations. Early approaches to scoring a candidate text with respect to a reference text were based on the idea that the similarity score should be proportional to the number of matching words (Melamed, 1995). Another idea is that matching words in the right order should result in higher scores than matching words out of order (Brew and Thompson, 1994; Rajman and Hartley, 2001).

As follows, the most widely-used MT evaluation metric, BLEU, is introduced. Other measures, such as NIST, mWER and mPER, are just referenced.

2.6.1.1   BLEU (Bilingual evaluation understudy)

Reported by (Papineni et al., 2002), the BLEU measure (acronym for BiLingual Evaluation Understudy) has dominated most machine translation work. Essentially, it consists of an N-gram corpus-level measure and it is always referred to as a given N-gram order ($BLEU_n$, n

usually being 4).

BLEU heavily rewards large N-gram matches between the source and target (reference) translations. Despite being a useful characteristic, this can often unnecessarily penalize syntactically valid but slightly altered translations with low N-gram matches. It is specifically designed to perform the evaluation on a corpus level and can perform badly if used over isolated sentences.

$BLEU_n$ is defined as:

$$BLEU_n = exp\left(\frac{\sum\limits_{i=1}^{n} bleu_i}{n} + length\_penalty\right) \qquad (2.17)$$

where $bleu_i$ is are cumulative counts (updated sentence by sentence) referred to the whole evaluation corpus (test and reference sets) and $length\_penalty$ is global. Even though these matching counts are computed on a sentence-by-sentence basis, the final score is *not* computed as a cumulative score, i.e. it is not computed by accumulating a given sentence score.

Equations 2.18 and 2.19 show $bleu_n$ and $length\_penalty$ definitions, respectively:

$$bleu_n = log\left(\frac{Nmatched_n}{Ntest_n}\right) \qquad (2.18)$$

$$length\_penalty = min\left\{0, 1 - \frac{shortest\_ref\_length}{Ntest_1}\right\} \qquad (2.19)$$

Finally, $Nmatched_i$, $Ntest_i$ and $shortest\_ref\_length$ are also cumulative counts (updated sentence by sentence), defined as:

$$Nmatched_i = \sum_{n=1}^{N} \sum_{ngr \in S} min\left\{N(test_n, ngr), \max_r \{N(ref_{n,r}, ngr)\}\right\} \qquad (2.20)$$

where S is the set of N-grams of size i in sentence $test_n$, $N(sent, ngr)$ is the number of occurrences of the N-gram ngr in sentence $sent$, N is the number of sentences to eval, $test_i$ is the ith sentence of the test set, R is the number of different references for each test sentence and $ref_{n,r}$ is the rth reference of the nth test sentence.

$$Ntest_i = \sum_{n=1}^{N} length(test_n) - i + 1 \tag{2.21}$$

$$shortest\_ref\_length = \sum_{n=1}^{N} \min_r \{length(ref_{n,r})\} \tag{2.22}$$

From the BLEU description, we can conclude:

- BLEU is a quality metric and it is defined in a range between 0 and 1, 0 meaning the worst translation (which does not match the references in any word), and 1 the *perfect* translation.

- BLEU is mostly a measure of *precision*, as $bleu_n$ is computed by dividing by the matching n-grams by the number of n-grams in the test (*not* in the reference). In this sense, a very high BLEU could be achieved with a *short* output, so long as all its n-grams are present in a reference.

- The *recall* or *coverage* effect is weighted through the *length_penalty*. However, this is a very rough approach to recall, as it only takes lengths into account.

- Finally, the weight of each effect (precision and recall) might not be clear, being very difficult from a given BLEU score to know whether the provided translation lacks recall, precision or both.

Although BLEU has been extensively used for comparative evaluation of the various MT systems developed by MT researchers, slight variations of these definitions have led to alternative versions of the BLEU score. Very recently, an interesting discussion with counterexamples of human correlation was presented in (Callison-Burch et al., 2006).

2.6.1.2   Other automatic measures

Apart from these, several other automatic evaluation measures that compare hypothesis translations against supplied references have been introduced, claiming strong correlation with human intuition. Although not used in this PhD dissertation, here we refer to some of them.

- NIST (Doddington, 2002) is an accuracy measure that calculates how informative a particular *n*-gram is, and the rarer a correct *n*-gram is, the more weight it will be given. Small variations in translation length do not impact much in the overall score.

- Word Error Rate (WER) (McCowan et al., 2004) is a standard speech recognition evaluation metric. One general difficulty of measuring performance lies in the fact that the translated word sequence can have a different length from the reference word sequence (supposedly the correct one). The WER is derived from the Levenshtein distance, working at the word level. Additionally for translation, its multiple-reference version (mWER) is computed on a sentence-by-sentence basis, so that the final measure for a given corpus is based on the cumulative WER for each sentence. Similar to WER, there is the Position-Independent Error Rate (mPER) which is again computed on a sentence-by-sentence basis. The main difference with WER is that it does not penalize the wrong order in the translation.

- Metric for Evaluation of Translation with Explicit ORdering (**METEOR**) evaluates a translation by computing a score based on explicit word-to-word matches between the translation and a reference translation. If more than one reference translation is available, the given translation is scored against each reference independently, and the best score is reported. It has been found (Lavie and Agarwal, 2007) that it correlates well with human adequacy.

- From a more intuitive point of view, in (Snover et al., 2005) Translation Error Rate, or **TER**, is presented. This measures the amount of editing that a human would have to perform to change a system output so it exactly matches a reference translation. Its application in real-life situation is reported in (Przybocki et al., 2006).

- The IQMT framework is presented in (Giménez and Amigó, 2006). This tool follows a 'divide and conquer' strategy, so that one can define a set of metrics and then combine them into a single measure of MT quality in a robust and elegant manner, avoiding scaling problems and metric weightings.

several automatic measures were presented and evaluated, and the ones that correlated better with human judgments were:

- Semantic role overlap (Giménez and Màrquez, 2007). This metric calculates the lexical overlap between semantic roles (i.e., semantic arguments or adjuncts) of the same type in the hypothesis and reference translations. It uniformly averages lexical overlap over all semantic role types.

- Dependency overlap (Amigó et al., 2006). This metric uses dependency trees for the hypothesis and reference translations, by computing the average overlap between words in the two trees which are dominated by grammatical relationships of the same type.

- ParaEval precision and ParaEval recall (Zhou et al., 2006). ParaEval matches hypothesis and reference translations using paraphrases that are extracted from parallel corpora in an unsupervised fashion. It calculates precision and recall using a unigram counting strategy.

MT evaluation has consequently been an area of significant research in itself over the years. A wide range of assessment measures have been proposed, not all of which are easily quantifiable. Recently developed frameworks, such as FEMTI (King et al., 2003), are attempting to devise effective platforms for combining multi-faceted measures for MT evaluation in effective and user-adjustable ways.

2.6.2   Human Evaluation

Evaluation of Machine Translation has traditionally been performed by humans. While the main criteria that should be taken into account in assessing the quality of MT output are fairly intuitive and well established, the overall task of MT evaluation is both complex and task dependent.

The most popular method is to obtain ratings from monolingual judges for segments of a document. The judges are presented with a segment, and are asked to rate it for two variables: adequacy and fluency. Adequacy is a rating of how much information is transferred between the original and the translation, and fluency is a rating of how good the English is. This technique is found to cover the relevant parts of the quality evaluation, while at the same time being easier to deploy, as it does not require expert judgment.

Measuring systems based on adequacy and fluency, along with informativeness is now the standard methodology for the ARPA evaluation program.

Among the different challenges in human evaluation we can name the following ones:

- Consistency. the evaluator himself and the other evaluators. An evaluator normally evaluates from 1 to 5 in fluency and adequacy each. It is difficult to maintain consistency in the own evaluations and at the same time, it is even more difficult to keep pace with others evaluators. Generally, this challenge is overcome by calculating an average among the evaluators.

- Learning from evaluation. Evaluating different types of errors would allow to distinguish system which perform more or less the same.

Another trend is to manually post-edit the references with information from the test hypothesis translations, so that differences between translation and reference account only for errors and the final score is not influenced by the effects of synonymy. The human targeted reference is obtained by editing the output with two main constraints, namely that the resultant references preserves the meaning and that it is fluent.

In this case, we refer to the measures as their human-targeted variants, such as HBLEU, HMETEOR or HTER as in (Snover et al., 2005). Unfortunately, this evaluation technique is also costly and cannot be used constantly to evaluate minor system improvements. Yet, we are of the opinion that, in the near future, these methods will gain popularity due to the fact that, apart from providing a well-founded absolute quality score, they produce new reference translations that can serve to automatically detect and classify translation errors.

Additionally, as a kind of MT evaluation, there are the international evaluation campaigns on MT. The goal of the international evaluation campaigns is to set up a framework for objectively comparing different MT systems. Through this, progress is to be promoted for MT technologies. Appendix § B overviews the most relevant MT international evaluation campaigns and it discusses the most significant aspects of our participation.

3

Continuous Space Language Models in Phrase- and Ngram-based SMT Systems

As stated in earlier chapters, the SMT framework formulates the problem of translating a sentence from a source language s into a target language t as the maximization problem of the conditional probability $p(t|s)$. During the translation process, a statistical score based on the probabilities of the feature functions is assigned to each translation candidate, and the one with the highest combination score is selected as translation output. However, the SMT system might not be able to correctly score translations due to statistical models limitations.

Introducing more complex feature functions that facilitate scoring the translation may not be easily introduced during decoding; for example, those feature functions that use the entire sentence to produce a score. One straightforward solution to this problem is a two-step decoding approach: in the first step, the decoder is run in n-best mode to produce n-best lists with N hypotheses per sentence. In the second step, the n-best lists are rescored with additional models. This technique reevaluates the n-best translation hypotheses of an MT system by introducing additional feature functions that should add information not included during decoding. Figure 3.1 shows a standard rescoring framework.

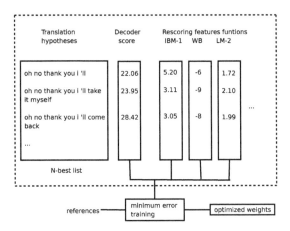

Figure 3.1: *Standard rescoring framework.*

This chapter [1] reports the research introducing continuous space language models (CSLM) during the rescoring step of phrase- and Ngram-based SMT systems. Section 3.1 reports the related work both in rescoring and language modeling applied to translation. Section 3.2 briefly describes continuous space language models that compute the feature function used for rescoring purposes, while Section 3.3 reports the influence of CSLM in rescoring the target language model both in a phrase- and Ngram-based systems. Section 3.4 reports the influence of CSLM in rescoring the translation language model in an Ngram-based system, which is the main contribution of this chapter. Finally, Section 3.5 concludes the chapter.

3.1   Motivation and Related Work

The first SMT studies in rescoring appeared with the goal of integrating feature functions in the log-linear SMT framework, which cannot be easily incorporated into a dynamic programming search. As stated before, one straightforward solution to this problem is to use a SMT decoder that computes a list of *n*-best translations and then rescore the list with the challenging feature function.

In (Och et al., 2004), rescoring allowed different syntactical features to be assessed quickly

---

[1]This work has been done in cooperation with Holger Schwenk from LIUM (University of Maine, France) who directly provided the Continuous Space Language Model.

without changing the search algorithm. More than 450 different feature functions were used in order to improve the syntactic form of the MT output. By reranking the 1000-best list generated by the baseline MT system from (Och, 2003), the BLEU score was improved from 31.6% to 32.9%.

A widely known work is (Kumar and Byrne, 2004), which presents the Minimum Bayes-risk approach in rescoring. It aims to minimize expected loss of translation errors under loss functions that measure translation performance.

Later, clustered language models were implemented in (Hasan and Ney, 2005). They reported a language model based on clusters obtained by applying regular expressions to the training data and thus discriminating several different sentence types as, e.g. interrogatives, imperatives or enumerations. The main motivation rested on the observation that different sentence types also have different syntactic structures, and thus yield a varying distribution of n-grams reflecting their word order. These language models were introduced in the SMT system using an n-best rescoring experiment.

Here, we propose to introduce continuous space language models as a feature function in rescoring.

First, this continuous space LM uses the limited resources to their greatest advantage when building a monolingual language model (trained on the target language). Most of the SMT systems use an n-gram, back-off LM to train this target language model. Some sites have reported improvements using 5-gram word or class-based LMs (Paul et al., 2005; Menezes and Quirk, 2005), or even 9-gram prefix and suffix LMs (Tsukada et al., 2005). Language model adaptation was investigated in (Hewavitharana et al., 2005). Other interesting approaches include factored (Kirchhoff and Yang, 2005) or syntax-based language models (Charniak et al., 2003).

Second, CSLM aims at smoothing the bilingual Ngram language model. There have been some works that address smoothing techniques in phrase-based systems (Foster et al., 2006). Two types of phrase-table smoothing were compared: black-box and glass-box methods. Black-methods do not look inside phrases but instead treat them as atomic objects. With this assumption, all the methods developed for language modeling can be used. Glass-box methods decompose $P(\tilde{s}|\tilde{t})$ and/or $P(\tilde{t}|\tilde{s})$ into a set of lexical distributions. For instance, IBM-1 probabilities (Och et al., 2004), or other lexical translation probabilities (Koehn et al., 2003; Zens and Ney, 2004) have been suggested. Some form of glass-box smoothing is now used in all state-of-the-art statistical machine translation systems.

In an Ngram-based system, smoothing is obtained using the standard techniques developed for language modeling. Smoothing has been extensively investigated in the area of monolingual language modeling. A systematic comparison can be found, for instance, in (Chen and Goodman, 1999). Language models and phrase tables have in common that the probabilities of rare events may be overestimated. However, in language modeling probability mass must be redistributed in order to account for the unseen n-grams.

3.2   Continuous Space Language Models

The basic idea of continuous space language models, also called neural network language models, is to project the word indices onto a continuous space and to use a probability estimator operating on this space. Since the resulting probability functions are smooth functions of the word representation, better generalization to unknown n-grams can be expected. This is believed to be particularly important for tasks with limited resources. A neural network can be used to simultaneously learn the projection of the words onto the continuous space and to estimate the n-gram probabilities. This is still an Ngram approach, but the LM conditional probabilities are *interpolated* for any possible context of length n-1, instead of backing-off to shorter contexts. This approach was successfully used in large vocabulary continuous speech recognition (Schwenk, 2007), and initial experiments have shown that it can be used to improve a word-based statistical machine translation system (Schwenk et al., 2006). Here, the continuous space LM is applied to a state-of-the-art phrase and Ngram-based SMT system.

Continuous space language modeling has a much higher complexity than back-off n-gram modeling. However, the use of the continuous space LM can be heavily optimized when rescoring n-best lists, i.e. by grouping together all calls with the same context from the whole n-best list, resulting in only one forward pass through the neural network.

The architecture of CSLM is shown in Figure 3.2. A standard, fully-connected multi-layer perceptron is used. The inputs to the neural network are the indices of the $n-1$ previous words in the vocabulary

$$h_j = w_{j-n+1}, \ \ldots, w_{j-2}, w_{j-1}$$

and the outputs are the posterior probabilities of *all* words of the vocabulary:

$$P(w_j = i|h_j) \qquad \forall i \in [1, N] \tag{3.1}$$

where N is the size of the vocabulary. The input uses 1-of-n coding, i.e., the ith word of the vocabulary is coded by setting the ith element of the vector to 1 and all the other elements to 0. The ith line of the $N \times P$ dimensional shared projection matrix corresponds to the continuous representation of the ith word. Let us denote c_l as the projections, d_j as the hidden layer activities, o_i as the outputs, p_i as their softmax normalization, and m_{jl}, $l = 1 \ldots P(n-1)$, b_j, v_{ij} and k_i as the hidden and output layer weights and the corresponding biases, respectively. Using these notations, the neural network performs the following operations:

$$d_j = \tanh\left(\sum_{l=1}^{P(n-1)} m_{jl} c_l + b_j\right) \quad j = 1 \ldots H \tag{3.2}$$

$$o_i = \sum_{j=1}^{H} v_{ij} d_j + k_i \quad i = 1 \ldots N \tag{3.3}$$

$$p_i = t^{o_i} / \sum_{r=1}^{N} t^{o_r} \tag{3.4}$$

The value of the output neuron p_i corresponds directly to the probability $P(w_j = i|h_j)$. Training is performed with the standard back-propagation algorithm minimizing the following error function:

$$E = \sum_{i=1}^{N} t_i \log p_i + \beta\left(\sum_{jl} m_{jl}^2 + \sum_{ij} v_{ij}^2\right) \tag{3.5}$$

where t_i denotes the desired output, i.e., the probability should be 1.0 for the next word in the training sentence and 0.0 for all the other ones. The first part of this equation is the cross-entropy between the output and target probability distributions, and the second part is a regularization term that aims to prevent the neural network from over-fitting the training data (weight decay). The parameter β has to be determined experimentally. Training uses a re-sampling algorithm (Schwenk, 2007).

It can be shown that the outputs of a neural network trained in this manner converge to the posterior probabilities. Therefore, the neural network directly minimizes the perplexity on the training data. In addition, the gradient is back-propagated through the projection-layer, which means that the neural network learns the projection of the words onto the continuous space that

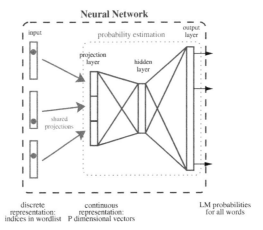

Figure 3.2: Architecture of the continuous space LM. h_j denotes the context $w_{j-n+1}, \ldots, w_{j-1}$. P is the size of one projection and H, N is the size of the hidden and output layer respectively. When short-lists are used the size of the output layer is much smaller then the size of the vocabulary.

is best for the probability estimation task.

3.3 Using Continuous Space Language Models to Recompute a Target Model

In this section, we investigate whether continuous space language models can be used to improve phrase- and Ngram-based SMT systems. Here, CSLM is used as a target language model in rescoring. CSLM is incorporated into a phrase- and Ngram-based translation system using n-best lists. In all our experiments, the language model probabilities provided by CSLM are used to replace those of the default target language model used by the decoder.

Practice has shown that it is advantageous to interpolate the continuous space LM with the reference back-off LM, since both seem to be complementary. The interpolation coefficient is calculated by optimizing perplexity on the development data. This interpolation is used in all our experiments. For the sake of simplicity we will still call this interpolation the continuous space LM.

Calculating a LM probability with a back-off model basically corresponds to a look-up table using hashing techniques, while a forward pass through the neural network is necessary for the continuous space LM. More details can be found in (Schwenk, 2007).

3.3.1 Task and System Description

Translation of four different languages is considered: Mandarin to English (Zh2En), Japanese to English (Jp2En), Arabic to English (Ar2En) and Italian to English (It2En) using the 2006 IWSLT data (see Chapter § A.1) and the corresponding official evaluation test set. These languages exhibit very different characteristics, e.g. with respect to word order, which may affect the role of the target LM. Parameters of the baseline systems are summarized in Table 3.1. Experiments are reported in phrase and Ngram-based systems whose baseline systems are computed as follows.

	Phrase-based system
Phrase-length	10
max target per source	30
Language model	4gram kneser-ney interpolation
Features	s2t and t2s lexicon models, word and phrase bonus
Beam	50
Search	Non-Monotonic
Reordering	m=5 j=3

	Ngram-based system
Tuple-length	no limit
solving NULL	IBM-1 approach
max target per source	30
Language model	4gram kneser-ney interpolation
Features	s2t and t2s lexicon models, word bonus
Beam	50
Search	Non-Monotonic
Reordering	m=5 j=3

Table 3.1: *Phrase (left) and Ngram (right) default parameters.*

Language-dependent preprocessing. For all language pairs, training sentences were split by using full stops on both sides of the bilingual text (when the number of stops was equal), increasing the number of sentences and reducing their length. Specific preprocessing for each language is detailed in its respective section below.

Regarding English, standard tokenization techniques were used as preprocessing.

Considering Arabic, we use the Buckwalter Arabic Morphological Analyzer[2] to obtain possible word analysis. The Morphological Analysis and Disambiguation for Arabic (MADA) tool (Habash and Rambow, 2005), kindly provided by the University of Columbia was used to reduce ambiguities. Once analyzed, Arabic words were segmented by separating all prefixes (prepositions, conjunctions, the article and the future marker) and suffixes (pronominal clitics).

Chinese preprocessing included re-segmentation using ICTCLAS (Zhang et al., 2003).

Italian was POS-tagged and lemmatized using the freely-available FreeLing morpho-

---

[2]Version 2.0. Linguistic Data Consortium Catalog: LDC2004L02.

syntactic analysis package (Atserias et al., 2006). Additionally, Italian contracted prepositions were separated into preposition and article, for example 'alla'→'a la', 'degli'→'di gli' or 'dallo'→'da lo'.

When dealing with Japanese, one has to come up with new methods for overcoming the absence of delimiters between words. We addressed this issue by word segmentation using the freely available JUMAN tool (Matsumoto and Nagao, 1994) version 5.1.

Word alignment. Word alignment was computed using the GIZA++ tool (Och and Ney, 2003). We aligned both translation directions and combined the two alignments with the union operation.

Phrase modeling. Phrase sets for each translation direction were extracted from the union set of alignments. Phrases up to length 10 were considered. The conditional and posterior probability were used as translation models.

Tuple modeling. Tuple sets for each translation direction were extracted from the union set of alignments. The SRI language modeling toolkit (Stolcke, 2002)[3] was used to compute a 4-gram bilingual language model.

Feature functions. Lexicon models were used in the source-to-target and target-to-source directions. A word and phrase (the latter only for the phrase-based) bonus and a target language model were added in decoding. Kneser-Ney smoothing (Kneser and Ney, 1995) and interpolation of higher and lower n-grams were used to estimate for the target 4-gram language model.

Decoding. The decoder was set to perform histogram pruning, keeping the best $b = 50$ hypotheses (during optimization, histogram pruning is set to keep the best $b = 10$ hypotheses). When allowing for reordering in the SMT decoder, the $maxjumps$ distance-based reordering was used and the distortion limit (m) and reordering limit (j) (see § 5.1.1) were empirically set to $m = 5$ and $j = 3$, as they showed a good trade-off between quality and efficiency. No reordering was applied for the It2En task.

Optimization. Once models were computed, optimal log-linear coefficients were estimated for each translation direction and system configuration using an in-house implementation of the widely used downhill Simplex method (Nelder and Mead, 1965) (using the single-loop detailed in §2.5). Following the consensus strategy proposed in (Chen et al., 2005), the objective function was set to $100 \cdot BLEU + 4 \cdot NIST$.

CSLM. As in previous applications, the neural network was not used alone but interpolation

---

[3]http://www.speech.sri.com/projects/srilm/

was performed to combine several language models. First, the neural network and the reference back-off LM were interpolated together - this always improved performance since both seem to be complementary. Second, seven neural networks with different sizes of the continuous representation were trained and interpolated together.[4] This usually achieves better generalization behavior than training one larger neural network. The interpolation coefficients were calculated by optimizing perplexity on the development data, using an EM procedure. The obtained values were 0.33 for the back-off LM and about 0.1 for each CSLM. This interpolation was used in all our experiments. As mentioned previously, for the sake of simplicity we will still call this the continuous space LM.

An alternative would be to add a feature function and to combine all LMs under the log-linear model framework; using maximum BLEU training. This raises the interesting question of whether the two criteria (minimal perplexity versus maximal BLEU score) lead to equivalent performance when multiple language models are used in a SMT system. In previous experiments with a word-based statistical machine translation system, both approaches yielded similar performance (Schwenk et al., 2006).

In general, the complexity required to calculate one probability with the continuous space LM is dominated by the dimension of the output layer, since the size of the vocabulary (up to 200k) is usually much larger than the dimension of the hidden layer (300...600). Therefore, in previous applications of the continuous space LM, the output was limited to the s most frequent words, where s ranges between 2k and 12k (Schwenk, 2007). This was not necessary for the BTEC task since the entire training corpus contains less than 10k different words. Thus, this was the first time that the continuous space LM was used to predict the LM probabilities of all words in the vocabulary.

Each network was trained independently using early stopping on development data. Convergence was achieved after about 10 iterations through the training data. The other parameters are as follows:

- Context of three words (4-gram);

- The dimension of the continuous representations of the words were $P = 100, 120, 140, 150, 160, 180$ and 200;

- The dimension of the hidden layer was set to $H = 200$;

- The initial learning rate was 0.005 with an exponential decay, and;

---

[4] We did not try to find the minimal number of neural networks to be combined. It is quite likely that the interpolation of fewer networks would result in the same BLEU scores.

	Phrase-based system			Ngram-based system		
	Oracle	Ref.	CSLM	Oracle	Ref.	CSLM
Mandarin	33.1	20.68	21.97	32.0	20.84	21.83
Japanese	26.9	17.29	18.27	28.6	18.34	19.77
Arabic	40.1	27.92	30.28	41.6	29.09	30.89
Italian	56.2	41.66	44.03	58.1	41.65	44.67

Table 3.2: *BLEU scores on the development data. Oracle uses a back-off LM trained on the references, Ref. is the default system as submitted in the official IWSLT evaluation, and CSLM uses the continuous space language model.*

- The weight decay coefficient was set to $\beta = 0.00005$.

Perplexity on development data is a popular and easy measure to evaluate the quality of a language model. However, it is not clear if perplexity is a good criterion to predict improvements when the language model will be used in a SMT system. All seven reference translations were concatenated to one development corpus of 40k words. The perplexity of the 4-gram reference back-off LM on this data is 124.0. This could be reduced to 96.6 using the continuous space LM.

3.3.2 Result Analysis

3.3.2.1 Development Set

Table 3.2 gives the BLEU scores when the back-off and continuous space LM is used with phrase- and Ngram-based SMT systems. It is often informative to have an idea of the oracle BLEU score of the n-best lists. This was estimated by rescoring the n-best lists with a cheating back-off LM that was built on the concatenated seven reference translation.

The continuous space LM always achieved important BLEU score improvements, ranging from 1 point absolute (phrase-based system for Japanese) to 3 points absolute (Ngram-based system for Italian). On average the development gain is slightly higher in the Ngram-based system than for the phrase-based system.

The performance of the continuous space LM differs sensibly for the different translation directions. It is not surprising that the best improvements were obtained for the translation from Italian to English, two languages that are quite similar with respect to word order and the only task where reordering was not applied. The gain brought by the continuous space LM is 2.4 and 3 BLEU points for the phrase- and Ngram-based systems, respectively. For the two

Asian languages Mandarin and Japanese, whose sentence structure more drastically different from English word order, the improvement brought by the new LM is about 1 point BLEU. Surprisingly in Mandaring and Japanese, the hypotheses in the *n*-best lists differ mostly in the choice of words and phrases, but there is less variation in the word order.

	Phrase-based			Ngram-based		
	Oracle	Ref.	CSLM	Oracle	Ref.	CSLM
Zh2En						
mWER	59.1	67.4	66.5	58.1	67.8	66.6
mPER	44.5	50.8	50.1	45.3	51.5	50.6
Jp2En						
mWER	70.8	74.6	77.0	63.5	73.0	71.3
mPER	48.5	52.2	54.6	46.1	53.4	51.8
Ar2En						
mWER	49.1	56.0	52.7	48.1	55.7	52.8
mPER	40.3	45.7	43.3	39.6	44.0	42.5
It2En						
mWER	34.1	42.3	40.7	33.1	42.8	40.8
mPER	26.6	31.6	30.5	26.0	31.9	30.7

Table 3.3: *Word error rates on the development data.*

The (position independent) word error rates are given in Table 3.3. The most interesting case is the translation from Arabic to English: the word error rate decreases 3.3 points and the position independent word error rate 2.4 points. There are also important word error reductions for the translation direction Japanese to English when using the Ngram SMT system. We cannot explain why the error rates increase for the phrase-based system. Moreover, the absolute values are much higher for this language. One can also notice that the position independent oracle word error rates are only about 5 percentage points lower than those of the best system. This may indicate that the *n*-best list generation could be improved in order to include more alternative translations.

In addition to the automatic scores we give some example translations in Figure 3.3. It seems that the continuous space LM manages to improve the fluency of the translation in some cases, for instance the phrase *we arrive time is two thirty* is replaced by *we arrive at two thirty*, *two and the fifty minutes* by *two and fifty minutes*, *two o'clock and thirty* by *two thirty*, *you ask* by *you can ask* and *are very busy* by *I'm very busy*. Although the meaning seems to be pretty much preserved in the translations from Asian languages, fluency is clearly worse than in It2En.

Reference translations:

ref 1:   *for your information we will arrive at two thirty and your departure time is at two fifty*
          oh sorry i'm very busy now you can ask someone else

ref 2:   *please refer to this information arrival at two thirty and departure at two fifty*
          sorry i'm a bit tied up at this moment could you ask someone else

ref 3:   *for your reference landing time is two thirty and take off time is two fifty*
          i'm sorry i can't help since i'm busy now please try someone else

ref 4:   *we will arrive at two thirty and set out at two fifty please consider this information*
          sorry i'm a bit too busy right now it'll be nice if you'd ask someone else

ref 5:   *you will need to consider that we will arrive at two thirty and your next flight is at two fifty*
          i'm sorry i'm busy at the moment please ask someone else

ref 6:   *please bear in mind that we touch down at two thirty and your connecting flight is at two fifty*
          i'm sorry but i'm very busy now you can try asking someone else

ref 7:   *please be mindful that we arrive at two thirty and your next departure is at two fifty*
          sorry but i'm very busy now you might want to ask someone else

---

Phrase-based system:

Zh2En baseline: *could you we arrive time is two thirty departure time is two five ten*
                   oh i'm sorry but i'm busy right now you can ask someone else

       cslm: *you can the time we arrive at two thirty departure time is two fifty*
                   oh i'm sorry but i'm busy right now you can ask someone else please

Jp2En baseline: *we arrive at two thirty take off time at two fifty in your it you please*
                   i'm sorry but my hand you had better ask someone else you can

       cslm: *we arrive at two thirty take off schedule at two fifty in you it you please*
                   i'm sorry my hand you had better ask someone else i can do

Ar2En baseline: *information your will we arrive at two thirty and an appointment is two and the fifty minutes*
                   excuse me i'm very busy 's ask someone else

       cslm: *information i'll arrive at two thirty and time is two and fifty minutes*
                   excuse me i'm very busy's ask someone else

It2En baseline: *for your information we'll be arriving at two o'clock and thirty and your departure time is at two o'clock and fifty*
                   oh sorry i'm very busy right now you can ask someone else

       cslm: *for your information we'll arrive at two thirty and your departure time is at two fifty*
                   oh sorry i'm very busy now you can ask someone else

---

Ngram-based system:

Zh2En baseline: *you can reference our arrival time is two thirty departure time is two fifty*
                   oh i'm sorry i'm in a hurry you ask someone else

       cslm: *you can reference our arrival time is two thirty departure time is two fifty*
                   oh i'm sorry i'm busy right now you can ask someone else

Jp2En baseline: *we arrive at two thirty takeoff time is fifty two o'clock so you reference you please*
                   i'm sorry but i can't get them eyes someone else to ask you can

       cslm: *we arrive at two thirty take off time is two o'clock in fifty so you your reference please*
                   i'm sorry but i can't get them eyes someone else ask you can

Ar2En baseline: *i'll information you arrive at two thirty time and is two and fifty minutes*
                   excuse me i'm very busy it's ask someone else

       cslm: *i'll information you arrive at two thirty and time is two and fifty minutes*
                   excuse me i'm very busy it's ask someone else

It2En baseline: *it's for your information we'll be arriving at two thirty and your departure time is at two fifty*
                   oh excuse me are very busy right now you can ask someone else

       cslm: *it's for your information we'll arrive at two thirty and your departure time is at two fifty*
                   oh sorry i'm very busy right now you can ask someone else

---

Figure 3.3: *Example translations using the baseline back-off and the continuous space language model (CSLM).*

3.3.2.2 Test Set

The results on the official test data of the 2006 IWSLT evaluation are summarized in Table 3.4. The numbers in the columns *reference* corresponds to the official results of the phrase and Ngram-based SMT systems that UPC has developed. The numbers in the columns *CSLM* were obtained by rescoring the *n*-best lists of these official systems with the continuous space language model described in this paper, using the coefficients of the feature functions that were tuned on the development data.

As usual, the improvements on the test data are smaller than those on the development data which was used to tune the parameters. As a rule of thumb, the gain on the test data is often half as large as on the development data. It seems that the phrase-based system generalizes slightly better than the Ngram-based approach: there is less difference between the improvements on the development and those on the test data. This does not seem to be related to the use of continuous space LMs since such behavior was previously observed on other tasks. One possibility could be that the Ngram approach is more sensitive to over-fitting when tuning the feature function coefficients. However, the Ngram-based SMT systems still achieve better BLEU scores than the phrase-based systems in three out of four tasks.

Another surprising result is the bad performance of the continuous space language model for the translation of Arabic to English with the Ngram-based system: the BLEU and NIST scores decrease despite an improvement in the word error rates. In fact, the reference Ngram-based SMT system does not perform very well on the test data in comparison to the phrase-based system. While its BLEU score was almost 1.2 points better on the development data it achieves basically the same result on the test data (23.83 BLEU with respect to 23.72).

3.4 Using Continuous Space Language Models to Re-compute the Translation Model of the Ngram-based System

In this section, we explore whether the continuous space LM can be successfully used to rescore the translation model of the Ngram-based system. This results in a continuous space translation model (CSTM).

Here, we investigate whether the continuous space language modeling is useful to smooth the probabilities involved in the bilingual tuple translation model. Reliable estimation of unseen

	Phrase-based		Ngram-based	
	Ref.	CSLM	Ref.	CSLM
Zh2En				
BLEU	19.74	**21.01**	20.34	**21.16**
NIST	6.24	**6.55**	6.22	**6.40**
mWER	**67.95**	68.16	68.30	**67.63**
mPER	52.46	**51.87**	52.81	**52.31**
Jp2En				
BLEU	15.11	**15.73**	16.14	**16.35**
NIST	5.83	**5.99**	5.86	**5.87**
mWER	**77.51**	78.15	**75.45**	75.59
mPER	55.14	**54.96**	55.52	**55.29**
Ar2En				
BLEU	23.72	**24.86**	**23.83**	23.70
NIST	**6.72**	6.69	**6.80**	6.70
mWER	63.04	**60.89**	62.81	**61.97**
mPER	49.43	**48.61**	49.41	**48.85**
It2En				
BLEU	35.55	**37.41**	35.95	**37.65**
NIST	8.32	**8.53**	8.40	**8.57**
mWER	49.12	**47.22**	48.78	**47.59**
mPER	38.17	**36.62**	38.12	**37.26**

Table 3.4: *Result summary of the test data.*

n-grams is very important in this translation model. Most of the trigram tuples encountered in the development or test data were never seen in the training data. N-gram hit rates are reported.

A big challenge was to introduce the CSTM in an Ngram-based system. Similarly to § 3.3, we used an n-best list. However, the main difference is that each n-best hypothesis had to be composed of a sequence of bilingual tuples, which corresponded to the CSTM training. Additionally, each hypothesis had the corresponding scores of all the feature functions. Figure 3.4 shows an example of such an n-best list. The CSTM was used to replace the scores of the back-off translation model. This was followed by re-optimization of the coefficients of all feature functions.

3.4.1   Task and System Description

Translation from Italian to English using the 2006 IWSLT data A.1 is considered. Experimental results are reported in an Ngram-based system built as described in §3.3.1, except for the non-

spiacente#sorry tutto_occupato#it_'s_full
spiacente#i_'m_sorry tutto_occupato#it_'s_full
spiacente#i_'m_afraid tutto_occupato#it_'s_full
spiacente#sorry tutto#all occupato#busy
spiacente#sorry tutto#all occupato#taken

Figure 3.4: *Example of sentences in the n-best list of bilingual tuples. '#' is used to separate the source and target sentence words. '_' is used to group several words in one tuple.*

monotonic decoding, with local reordering ($m = 5$ and $j = 3$). Parameters of the baseline system are summarized in Table 3.1.

The training details for the continuous space LM are also kept as in §3.3.1. The training details for the continuous space translation model are described as follows.

CSTM. As explained for an Ngram-based baseline system § 2.3.2, given the alignment of the training parallel corpus, we performed a unique segmentation of each parallel sentence using the unfolding technique. This segmentation was used in a sequence as training text for building the language model.

The reference bilingual trigram back-off translation model was trained on these bilingual tuples using the SRILM toolkit (Stolcke, 2002). Different smoothing techniques were tried, and the best results here were obtained using Good-Turing discounting.

The continuous space approach was trained on exactly the same data. A context of two tuples was used (trigram model). The training corpus contained about 21,500 different bilingual tuples. We decided to limit the output of the neural network to the 8k most frequent tuples (short-list). This covered about 90% of the requested tuple n-grams in the training data.

Similar to the continuous space LM in § 3.3, the neural network was not used alone, but rather interpolation was performed to combine several n-gram models. First, the continuous space and the reference back-off models were interpolated together - this always improved performance since they seem to be complementary. Second, four neural networks with different sizes of the continuous representation were trained and interpolated together. This usually achieves better generalization behavior than training one larger neural network. The interpolation coefficients were calculated by optimizing perplexity on the development data, using an EM procedure. The obtained values were 0.33 for the back-off translation model and about 0.16 for each continuous space model. This interpolation was used in all our experiments. For the sake of simplicity we will still call this the continuous space translation model.

Each network was trained independently using early stopping on the development data. Convergence was achieved after about 10 iterations through the training data. The other parameters are as follows:

- Context of two tuples (trigram);

- The dimension of the continuous representation of the tuples were P =120,140,150 and 200;

- The dimension of the hidden layer was set to $H = 200$;

- The initial learning rate was 0.005 with an exponential decay; and

- The weight decay coefficient was set to $\beta = 0.00005$.

As stated before, N-gram models are usually evaluated using perplexity on some develop-ment data. In our case, i.e. using bilingual tuples as basic units (words), it is less obvious if perplexity is a useful measure. Nevertheless, we provide these numbers for completeness. The perplexity on the development data of the trigram back-off translation model is 227.0. This could be reduced to 170.4 using the neural network. It is also very informative to analyze the n-gram hit-rates of the back-off model: 8% of the probability requests are actually a true trigram, 32% a bigram and about 60% are finally estimated using unigram probabilities. This means that only a limited amount of phrase context is used in the standard Ngram-based translation model. This makes it an ideal candidate for the continuous space model since probabilities are interpolated for all possible contexts and never backed-up to shorter contexts.

In all our experiments 1000-best lists were used and they were optimized to maximize the BLEU score on the development data using the numerical optimization method CONDOR (Berghen and Bersini, 2005).

3.4.2   Result Analysis

In order to evaluate the quality of these n-best lists, an oracle trigram back-off translation model was build on the development data. Rescoring the n-best lists with this translation model re-sulted in an increase of the BLEU score by about 10 points (see Table 3.5). It is worth noticing that, while there is a decrease of about 6% for the position dependent word error rate (mWER), a smaller change in the position independent word error rate was observed (mPER). This sug-gests that there may be more alternative translation hypothesis in word reorderings than in word choices. This happens in It2En where there are mainly local reorderings, the n-best list offers several reordered hypotheses and the CSLM rescoring is capable of choosing the best alternative among them.

	Back-off	Oracle	CSLM
BLEU	42.34	52.45	**43.87**
mWER	41.6%	35.6%	**40.3%**
mPER	31.5%	28.2%	**30.7%**

Table 3.5: *Comparison of different N-gram-translation models of the development data.*

When the 1000-best lists are rescored with the continuous space translation model the BLEU score increases by 1.5 points (42.34 to 43.87). Similar improvements were observed in the word error rates (see Table 3.5). For comparison, a 4-gram back-off translation model was also built, but no change of the BLEU score was observed. This suggests that careful smoothing is more important than increasing the context when estimating the translation probabilities in an Ngram-based SMT system.

As shown in the earlier subsection, the use of the continuous space approach to modeling the target language was investigated for the IWSLT task. We also applied this technique to our Ngram-based translation system. In our implementation, the continuous space target 4-gram language model gives an improvement of 1.3 points BLEU on the development data (42.34 to 43.66), in comparison to 1.5 points for the continuous space translation model see Table 3.6).

	Back-off TM+LM	CS TM	CS LM	CS TM+LM
BLEU	42.34	43.87	43.66	**44.83**

Table 3.6: *Combination of a continuous space translation model (TM) and a continuous language model (LM). BLEU scores on the development data.*

The continuous space translation and target language model were also applied to the test data, using the same feature function coefficients as for the development data. The results are given in Table 3.7 for all the official measures of the IWSLT evaluation. The new smoothing method of the translation probabilities achieves improvement in all measures. Additionally, it gives additional gain in all measures when used together with a continuous space target language model. Surprisingly, continuous space TM and LM improvements are almost completely additive: when both techniques are used together, the BLEU scores increases by 1.5 points (36.97 → 38.50).

	Back-off TM+LM	CS TM	CS LM	CS TM+LM
BLEU	36.97	37.21	38.04	**38.50**
mWER	48.10	**47.42**	47.83	47.61
mPER	38.21	38.07	37.26	**37.12**
NIST	8.3	8.3	8.6	**8.7**
Meteor	63.16	63.40	64.70	**65.20**

Table 3.7: *Combination of a continuous space translation model (TM) and a continuous space language model (LM). BLEU scores on the test data.*

3.5   Conclusions and Further Work

The use of a continuous space LM performs probability estimation in a continuous space. Since the resulting probability functions are smooth functions of the word representation, better generalization to unknown *n*-grams can be expected. This is particularly interesting for tasks like the BTEC corpus where only limited amounts of appropriate LM training material are available.

The continuous space LM is introduced as a target language model to rescore the *n*-best lists of a phrase- and Ngram-based statistical machine translation system. We have studied 4-gram language models because we had limited data. With more data, it would be easy to train a continuous space LM with much longer contexts, since the complexity of our approach increases only slightly with the size of the context. Results are provided on the BTEC tasks of the 2006 IWSLT evaluation for the translation direction Chinese, Arabic, Japanese and Italian to English. These tasks provide a very limited amount of resources in comparison to other tasks. Therefore, new techniques must be employed to take the best advantage of limited resources. The results show significant improvement for four different languages pairs and for both systems. The new approach achieves good improvements on the test data; the BLEU score increases by up to 1.9 points. All these results were obtained on top of the evaluation systems.

The continuous space language modeling was successfully extended to smoothing the bilingual language model of an Ngram-based system §3.4. The continuous space language model is trained on a bilingual sequence of tuples and it is introduced in the Ngram-based system rescoring. Our method is distinguished by two characteristics: better estimation of the numerous unseen *n*-grams; and a discriminative estimation of the tuple probabilities. Results are provided on the BTEC task of the 2006 IWSLT evaluation for the translation direction Italian to English. We have chosen the Italian to English task because it is challenging to improve the already good quality of the translation task (over 40 BLEU). Using the neural model for the translation and target language model, an improvement of 1.5 BLEU points on the test data was observed.

The described smoothing method was explicitly developed to tackle the data sparseness problem in tasks like the BTEC corpus. Recently, continuous space language modeling applied on the target model (of a phrase-based system) has shown significant improvements when large amounts of data are available (LIUM site reported 1 point BLEU of improvement in the 2008 NIST evaluation). As future work, we plan to investigate whether this also holds in an Ngram-based statistical machine translation system where both the translation and target language model can be rescored with the continuous space language model.

RELATED PUBLICATIONS

- Schwenk, H., Costa-jussà, M.R. and Fonollosa, J.A.R.
  Continuous Space Language Models for the IWSLT 2006 task
  International Workshop on Spoken Language Translation (IWSLT) pages 166-173 Kyoto, November 2006.

- Schwenk, H., Costa-jussà, M.R. and Fonollosa, J.A.R.
  Smooth Bilingual Translation
  Empirical Methods in Natural Language Processing (EMNLP) pages 430-438 Prague, June 2007.

4

Phrase and Ngram-based System Combination

Multiples translations can be computed by one MT system or by different MT systems. We may assume that different MT systems make different errors due to using different models, generation strategies or tweaks. An investigated technique, inherited from Automatic Speech Recognition, is the so-called system combination that is based on combining the outputs of multiples MT systems.

Given that most experiments of this PhD were done with phrase- and Ngram-based SMT systems, we combined the outputs of both systems using statistical criteria and additional rescoring features.

This chapter reports the research which has been carried out combining one and multiple translations from phrase- and Ngram-based systems. Section 4.1 reports the motivation of our experiments and the related work in system combination. Section 4.2 describes the straight two-system combination, and Section 4.3 presents a more elaborate system combination approach and a comparative analysis between systems. Finally, Section 4.4 concludes the chapter.

4.1   Motivation and Related Work

Combining outputs from different systems was shown to be quite successful in automatic speech recognition (ASR). Voting schemes like the ROVER approach of (Fiscus, 1997) use edit distance alignment and time information to create confusion networks from the output of several ASR systems.

In MT, some approaches combine lattices or n-best lists from several different MT systems (Frederking and Nirenburg, 1994). To be successful, such approaches require compatible lattices and comparable scores of the (word) hypotheses in the lattices.

(Bangalore et al., 2001) used the edit distance alignment extended to multiple sequences to construct a confusion network from several translation hypotheses. This algorithm produces monotone alignments only, i.e., allows insertion, deletion, and substitution of words. (Jayaraman and Lavie, 2005) try to deal with translation hypotheses with significantly different word order. They introduce a method that allows non-monotone alignments of words in different translation hypotheses for the same sentence.

Experiments combining several kinds of MT systems have been presented in (Matusov et al., 2006a), but they are only based on the single best output of each system. They propose an alignment procedure that explicitly models reordering of words in the hypotheses. In contrast to existing approaches, the context of the whole document, rather than a single sentence, is considered in this iterative, unsupervised procedure, yielding a more reliable alignment.

More recently, confusion networks have been generated by choosing one hypothesis as the skeleton and other hypotheses are aligned against it. The skeleton defines the word order of the combination output. Minimum Bayes Risk (MBR) was used to choose the skeleton in (Sim et al., 2007). The average translation edit rate (TER) score (Snover et al., 2006) was computed between each system's 1-best hypotheses in terms of TER. This work was extended by (Rosti et al., 2007) by introducing system weights for word confidences.

Finally, the most straightforward approach simply selects, for each sentence, one of the provided hypotheses. The selection is made based on the scores of translation, language, and other models (Nomoto, 2004; Doi et al., 2005).

In Section 4.2, we combine the output of two systems: phrase- and Ngram-based. Both systems are statistical and share similar features. That is why they tend to produce outputs which do not vary much. We propose a straightforward approach which simply selects, for each sentence, one of the hypotheses. In these systems, the phrase- or Ngram-based models are usually the main features in a log-linear framework, reminiscent of the maximum entropy modeling approach. Two basic issues differentiate both systems. In the Ngram-based model the training data is sequentially segmented into bilingual units, and the probability of these units is estimated as a bilingual Ngram language model. However, in the phrase-based model, no monotonicity restriction is imposed on the segmentation, and the probabilities are normally estimated simply by relative frequencies.

In Section 4.3, we propose to rescore the outputs of both systems with the probability given by both systems. The central point here is that the cost of phrase-based output is computed with the Ngram-based system. This cost is combined together with the phrase-based cost of this output. For the Ngram-based outputs, the cost is computed with the phrase-based system and again this cost is combined together with the Ngram-based cost. Notice that we have to deal with the particular cases where one translation generated by a system cannot be computed by the other system.

4.2   Straight System Combination

Integration of phrase and Ngram-based translation models in the search procedure would be a complex task. First, translation units of both models are quite different. Then, the fact that the Ngram-based translation model uses context and the phrase-based translation model does not use it poses severe implementation difficulties.

As we have seen in Chapter 3 some features that are useful for SMT are too complex to include directly in the search process. A clear example are the features that require the entire target sentence to be evaluated, as this is not compatible with the pruning and recombination procedures that are necessary for keeping the target sentence generation process manageable. A possible solution for this problem is to apply sentence level reranking by using the outputs of the systems, i.e. to carry out a two-step translation process (see § 4.3).

The aim of this preliminary system combination is to select the best translation given the 1-best output of each system (phrase- and Ngram-based) using the following feature functions:

- IBM-1 lexical parameters from the source to target direction and from the target to source direction. As seen in § 2.2, the IBM Model 1 is a word alignment model that is widely used in working with parallel corpora. Similarly to the feature function used in decoding (see § 2.4), the IBM-1 lexical parameters are used to estimate the translation probabilities of each hypothesis in the n-best list.

- Target language models (see § 2.1). Given that a 4-gram language model was used in decoding, we add a 2-gram, 3-gram and 5-gram. These models should be more useful when trained on additional monolingual data which is easier to acquire than bilingual data. Unfortunately, we were not able to add more data.

- Word bonus. Given that the above models tend to shorten the translation, translations receive a bonus for each produced word (see § 2.4).

4.2.1 Task and System Description

Translation of four different languages is considered: Mandarin to English, Japanese to English, Arabic to English and Italian to English using the 2006 IWSLT data (see Chapter A.1) and the corresponding official evaluation test set. Both phrase and Ngram-based systems are described in § 3.3.1, except for the reordering approach, which differs in some tasks, and it is described in (Costa-jussà et al., 2006; Crego et al., 2006b) for the phrase and Ngram-based system, respectively. Later in this thesis, we will describe these reordering approaches in Chapters 7 and 5.

The optimization tool used for computing each model weight both in decoding and rescoring was based on the simplex method (Nelder and Mead, 1965). Following the consensus strategy proposed in (Chen et al., 2005), the objective function was set to $100 \cdot BLEU + 4 \cdot NIST$.

Parameters of the baseline systems are summarized in Table 3.1.

4.2.2 Result Analysis

Table 4.1 shows that improvements due to system combination are consistent in the internal test set. Moreover, note that in the combined approach, a general improvement of the BLEU score is observed, whereas the NIST score seems to decrease. This behavior can be seen in almost all tasks and all test sets. Both the IBM-1 lexical parameters and the language model tend to benefit shorter outputs. Although a word bonus was used, we have seen that the outputs produced by the TALPcom system are shorter than the outputs produced by the TALPphr (phrase-based) or the TALPtup (Ngram-based) systems, which is why NIST did not improve.

When observing results in the test set, system combination performs well in all tasks improving from 2 to 3 point BLEU, except in the It2En task where BLEU stays the same.

Language	System	Dev		Test		Eval	
		BLEU	NIST	BLEU	NIST	BLEU	NIST
Zh2En	TALPphr primary	19.29	6.57	46.33	8.95	20.08	**6.42**
	TALPtup primary	19.75	6.64	44.63	**8.99**	**20.34**	6.22
	TALPcom	**21.19**	**6.69**	**49.72**	8.36	20.21	5.97
Ar2En	TALPphr primary	27.07	7.15	55.34	10.28	22.20	6.54
	TALPtup primary	29.27	**7.52**	55.11	10.45	23.83	**6.80**
	TALPcom	**30.29**	7.41	**57.34**	**10.46**	**23.95**	6.60
It2En	TALPphr primary	41.66	9.08	62.68	10.69	35.55	8.32
	TALPtup primary	43.05	**9.21**	**63.40**	**10.76**	37.38	**8.59**
	TALPcom	**44.13**	9.04	63.38	10.43	**37.74**	8.41
Jp2En	TALPphr primary	15.37	6.01	48.93	**9.54**	14.51	5.58
	TALPtup primary	16.59	6.34	47.14	9.42	14.61	5.27
	TALPcom	**19.36**	**6.42**	**51.73**	8.8	**15.66**	**5.51**

Table 4.1: *Results obtained using TALPphr (phrase-based), TALPtup (Ngram-based system) and their combination.*

When observing the results in the evaluation set, the analysis changes. We should take into account that the evaluation set from 2006 was not extracted from the BTEC Corpus, which may explain the difference in behavior. The biggest improvement is in Jp2En which leads to an improvement of 1 point BLEU. It2En follows, and is succeeded by Ar2En. Finally, Zh2En is the only case where the system combination decreases the translation quality in BLEU.

4.3 Phrase and Ngram-based Combination Score

We extend the method in § 4.2 by combining the two systems at the level of n-best lists doing a two-step translation.

In this case, the first step generates an n-best list using the models that can be computed at search time. In our approach, the translation candidates lists are produced independently by both systems and are then combined by concatenation[1]. During the second step, these multiple translation candidates are reranked using as additional feature function the probability given by the opposite system. Given the phrase-based (or Ngram-based) n-best list, we compute the cost of each target sentence of this n-best list given by the Ngram-based (or phrase-based) system. As a consequence, for each sentence we have the cost given by the phrase- and Ngram-based systems.

---

[1] With removal of duplicates.

However, this computation may not be possible in all cases. An example is given in Figure 4.1. Let us suppose that the three top sentences in Figure 4.1 are our bilingual training sentences with their corresponding word alignment. Then, the phrase and tuple extraction is computed following the criteria in § 2.3.2 and § 2.3.2, respectively. For the sake of simplicity, the phrase length was limited to 3. It is shown that several units are common in both systems and others are only in one system, which leads to different translation dictionaries. Finally, Figure 4.1 shows a test sentence and the corresponding translations of each system. Both sentences cannot be reproduced by the opposite system because there are target words that are not contained in the respective dictionaries given the actual source words. (the tuple dictionary does not contain the unit *translations # traducciones* and the phrase dictionary does not contain *They # NULL*).

Given the unique and monotonic segmentation of the Ngram-based system, the number of phrase-based translations that can be reproduced by the Ngram-based system may be smaller than the number of Ngram-based translations reproduced by the phrase-based system. Whenever a sentence cannot be reproduced by a given system, the cost of the worst sentence in the n-best list is assigned to it. This is an experimental decision for which we penalize sentences that cannot be produced by both systems.

Source: *We are about to achieve perfect translations* Target: *Estamos a punto de conseguir traducciones perfectas* Align: *1-0 2-1 3-2 4-2 4-3 5-4 7-5 6-6*	
Source: *They would like to achieve perfect translations* Target: *Ellos quieren lograr traducciones perfectas* Align: *1-0 2-1 3-1 4-2 5-2 6-4 7-3*	
Source: *They agree* Target: *Ellos están de acuerdo* Align: *1-1 2-2 2-3*	

TUPLE Extraction	PHRASE Extraction
We # NULL	
	We are # Estamos
	We are about # Estamos a punto
	are about # Estamos a punto
are # Estamos	
about # a punto	
	about to # a punto de
to # de	
	to achieve # de lograr
achieve # lograr	
traducciones perfectas # perfect translations	
	achieve perfect translations# lograr traducciones perfectas *perfect#perfectas* *translations#traducciones*
They # NULL	
	They would like # Quieren
They # Ellos	
agree # están de acuerdo	
would like # Quieren	
to achieve # lograr	
They are about to achieve incredible translations PHRASES: Ellos Estamos a punto de lograr incredible* *traducciones* TUPLES: *NULL* Estamos a punto de lograr incredible* *translations*	

Figure 4.1: *Analysis of phrase and Ngram-based systems. From top to bottom, training sentences with the corresponding alignment (source word position hyphen target word position), unit extraction, test sentence and translation were computed with each system using the extracted units. Word marked with * is unknown for the system.*

4.3.1   Task and System Description

The translation models used in the following experiments were used by UPC and RWTH in the second evaluation campaign of the TC-STAR project § A.3 (Spanish-English).

Preprocessing. Standard tools were used for tokenizing and filtering.

Word alignment. After preprocessing the training corpora, word-to-word alignments were performed in both alignment directions using Giza++ (Och and Ney, 2003) and the union set of both alignment directions was computed.

Tuple modeling. Tuple sets for each translation direction were extracted from the union set of alignments. The resulting tuple vocabularies were pruned out considering the N best translations for each tuple source-side ($N = 20$ for the Es2En direction and $N = 30$ for the En2Es direction) in terms of occurrences. The SRI language modeling toolkit (Stolcke, 2002)[2] was used to compute the bilingual 4-gram language model. Kneser-Ney smoothing (Kneser and Ney, 1995) and interpolation of higher and lower n-grams were used to estimate the 4-gram translation language models.

Phrase modeling. Phrase sets for each translation direction were extracted from the union set of alignments. The resulting phrase vocabularies were pruned out considering the N best translations for each tuple source-side ($N = 20$ for the Es2En direction and $N = 30$ for the En2Es direction) in terms of occurrences. Phrases up to length 10 were considered.

Feature functions. Lexicon models were used in the source-to-target and target-to-source directions. A word and a phrase (the latter only for the phrase-based system) bonus and a target language model were added in decoding. Again, Kneser-Ney smoothing (Kneser and Ney, 1995) and interpolation of higher and lower n-grams were used to estimate the 4-gram target language models.

Optimization. Once models were computed, optimal log-linear coefficients were estimated for each translation direction and system configuration using an in-house implementation of the widely used downhill Simplex method (Nelder and Mead, 1965) (using the single-loop detailed in §2.5). The BLEU score was used as the objective function.

Decoding. The decoder was set to perform histogram pruning, keeping the best $b = 50$ hypotheses (during the optimization work, histogram pruning is set to keep the best $b = 10$ hypotheses).

Parameters of the baseline systems are defined in Table 4.2.

---

[2]http://www.speech.sri.com/projects/srilm/

	Phrase-based system
Phrase-length	10
max target per source	20/30
Language model	4gram kneser-ney interpolation
Features	s2t and t2s lexicon models, word and phrase bonus
Beam	50
Search	Monotonic

	Ngram-based system
Tuple-length	no limit
solving NULL	IBM-1 approach
max target per source	20/30
Language model	4gram kneser-ney interpolation
Features	s2t and t2s lexicon models, word bonus
Beam	50
Search	Monotonic

Table 4.2: *Phrase (left) and Ngram (right) default parameters.*

4.3.2 Result Analysis

Table 4.3 shows results of the rescoring and system combination experiments on the development set. λ_{pb} is the weight given to the cost computed with the phrase-based system and λ_{nb} is the weight given to the cost computed with the Ngram-based system. For each translation direction, the first six rows include results of systems non-rescored and PB (NB) rescored by NB (PB). The last third rows correspond to the system combination where the PB (NB) outputs are concatenated with the NB (PB) outputs and ranked by their rescored score. The weight given to the Ngram score tends to be higher than the weight given to the phrase score, see *lambda*$_{pb}$ and *lambda*$_{nb}$ in Table 4.3. Moreover, the gain in rescoring is always higher for the phrase-based system. This improvement is enhanced by the number of *n*-best. Here, the best results are attained by the system combination.

First, we extracted the *n*-best list with the phrase-based system. Then we found the corresponding costs of this *n*-best list given by the Ngram-based system. Second, we extracted the *n*-best list with the Ngram-based system. Thirdly, we concatenated the first *n*-best lists with the second.

As explained in detail in § 4.3, given that the phrase- and Ngram-based system have different dictionaries, a phrase (Ngram) translation may not be reproduced by the opposite system. In that case, the cost of the worst sentence in the *n*-best list is assigned to it.

Rescoring the phrase-based system reaches an improvement of 0.8 point BLEU in the En2Es development set and a little less in the opposite direction. The Ngram-based baseline system is already better than the phrase-based baseline system and rescoring reaches an smaller improvement: 0.3 point BLEU in the development set in both directions. The improvement is much lower than in the case of rescoring a phrase-based system. Although, the *n*-best lists were more varied in the Ngram-based system, the quality of *n*-best translations may be higher in the phrase-based system. The system combination has almost reached 0.5 point BLEU of improvement when comparing it to the best of the two baseline systems and 1 point BLEU when comparing

System	n-best	BLEU	λ_{pb}	λ_{nb}
Es2En				
PB	1	45.45	1	-
PB	100	45.84	1	2.67
PB	1000	46.0	1	2.5
NB	1	46.01	-	1
NB	100	46.32	0.23	1
NB	1000	46.32	0.21	1
PB+NB	2	46.37	0.25	1
PB+NB	200	46.36	0.89	1
PB+NB	2000	**46.47**	0.18	1
En2Es				
PB	1	47.87	1	-
PB	100	48.57	1	1.96
PB	1000	48.66	1	1.16
NB	1	48.08	-	1
NB	100	48.3	0.50	1
NB	1000	48.35	1	1.16
PB+NB	2	48.32		1
PB+NB	200	48.51	0.55	1
PB+NB	2000	**48.54**	0.50	1

Table 4.3: *Results of rescoring and system combination on the development set. PB stands for phrase-based system and NB for Ngram-based system.*

it to the worst of the two baseline systems.

Table 4.4 shows the results of the rescoring and system combination experiments on the test set. Again, the first two rows include results of systems non-rescored and PB (NB) rescored by NB (PB). The third row corresponds to the system combination. Here, PB (NB) rescored by NB (PB) are simply merged and reranked.

Rescoring the phrase-based system reaches an improvement of 0.65 point BLEU in the test set. Rescoring the Ngram-based system reaches an improvement of 0.6 point BLEU, which is a better performance than in the development set. Finally, the Es2En system combination BLEU reaches 0.4 point BLEU improvement when comparing it to the best of the two baseline systems and 0.8 when comparing it to the worst system. In the opposite direction, the gain is about 0.8 point BLEU when comparing it to the two baseline systems. In En2Es, the improvement is consistent in all measures. In Es2En, the improvement in rescoring is coherent, but the improvement using the system combination is not. Here, the results in the development set do not generalize to the test set.

System	n-best	BLEU	NIST	mWER
Es2En				
PB	1	51.90	10.54	37.50
PB	100	52.45	10.59	37.17
PB	1000	**52.55**	**10.61**	**37.12**
NB	1	51.63	10.46	37.88
NB	100	52.16	10.54	37.53
NB	1000	52.25	10.55	37.43
PB+NB	2	51.77	10.49	37.68
PB+NB	200	52.12	10.54	37.58
PB+NB	2000	52.31	10.56	37.32
En2Es				
PB	1	47.75	9.94	41.2
PB	100	48.17	10.07	40.30
PB	1000	48.46	10.13	**39.98**
NB	1	47.73	10.09	40.50
NB	100	48.27	10.15	40.11
NB	1000	48.33	10.15	40.13
PB+NB	2	48.26	10.05	40.61
PB+NB	200	48.49	10.15	40.05
PB+NB	2000	**48.54**	**10.16**	**40.00**

Table 4.4: *Rescoring and system combination results.*

4.3.3   Structural Comparison

The following experiments were carried out to give a comparison of the translation units used in the phrase- and Ngram-based systems that were combined above.

Both approaches aim at improving accuracy by including word context in the model. However, the implementation of the models is quite different and may produce variations in several aspects. Table 4.5 shows how decoding time varies with the beam size. Additionally, the number of available translation units is shown, corresponding to the number of available phrases for the phrase-based system and 1gram, 2gram and 3gram entries for the Ngram-based system. Results are computed on the development set.

The number of translation units is similar in both tasks for both systems ($537k \sim 537k$ for Spanish-to-English and $594k \sim 651k$ for English-to-Spanish) while the time consumed during decoding is clearly higher for the phrase-based system. This can be explained by the fact that, in the phrase-based approach, the same translation can be hypothesized following several segmen-

Task	Beam	Time(s)	Units
	50	2,677	
Es2En	10	852	537k
	5	311	
	50	2,689	
En2Es	10	903	594k
	5	329	
	50	1,264	
Es2En	10	281	104k 288k 145k
	5	138	
	50	1,508	
En2Es	10	302	118k 355k 178k
	5	155	

Table 4.5: *Impact on efficiency of the beam size in PB (top) and NB system (bottom).*

tations of the input sentence, as phrases appear (and are collected) from multiple segmentations of the training sentence pairs. In other words, the search graph seems to be overpopulated in the phrase-based approach.

Table 4.6 shows the effect on translation accuracy regarding the size of the beam in the search. Results are computed on the test set for the phrase-based and Ngram-based systems.

Task	Beam	BLEU	NIST	mWER
	50	51.90	10.53	37.54
Es2En	10	51.93	10.54	37.49
	5	51.87	10.55	37.47
	50	47.75	9.94	41.20
En2Es	10	47.77	9.96	41.09
	5	47.86	10.00	40.74
	50	51.63	10.46	37.88
Es2En	10	51.50	10.45	37.83
	5	51.39	10.45	37.85
	50	47.73	10.08	40.50
En2Es	10	46.82	9.97	41.04
	5	45.59	9.83	41.04

Table 4.6: *Impact on accuracy of the beam size in PB (top) and NB system (bottom).*

Results of the Ngram-based system show that decreasing the beam size produces a clear reduction in the accuracy results. The phrase-based system shows that accuracy results remain very similar under the different settings. This is because of the way translation models are used in the search. In the phrase-based approach, every partial hypothesis is scored un-contextualized, hence, a single score is used for a given partial hypothesis (phrase). In the Ngram-based approach, the model is intrinsically contextualized, which means that each partial hypothesis (tuple) depends on the preceding sequence of tuples. Thus, if a bad sequence of tuples (poor scored) is composed of a good initial sequence (good scored), it is placed on top of the first stacks (beam) and may cause the pruning of the rest of hypotheses.

4.3.4 Error Analysis

This section aims at identifying the main problems in a phrase- and Ngram-based systems by performing a human error analysis. The main objective of this analysis is to focus on the differences between both systems, which justify the system combination. The guidelines for this error analysis can be found in (Vilar et al., 2006).

We randomly selected 100 sentences, to be evaluated by bilingual judges.This analysis reveals that the two systems produce some different errors (most are the same). For the English to Spanish direction, the greatest problem is the correct generation of the right tense for verbs, with around 20% of all translation errors being of this kind. Reordering also poses an important problem for both phrase and Ngram-based systems, with 18% or 15% (respectively) of the errors falling into this category. Missing words is also an important problem. However, most of them (approximately two thirds for both systems) are filler words (i.e. words which do not convey meaning). In general, the meaning of the sentence is preserved. The most remarkable difference when comparing both systems is that the Ngram based system produces a relatively large amount of extra words (approximately 10%), while for the phrase-based system, this is only a minor problem (2% of the errors). In contrast the phrase-based system has more problems with incorrect translations, that is words for which a human can find a correspondence in the source text, but the translation is incorrect.

Similar conclusions can be drawn for the inverse direction. The verb generation problem is not as acute in this translation direction due to the much simplified morphology of English.

An important problem is the generation of the correct preposition. The Ngram based system seems to be able to produce more accurate translations (reflected by a lower percentage of translation errors). However, it generates too many additional (and incorrect words) in the process. The phrase-based system, in contrast, counteracts this effect by producing a more direct correspondence with the words present in the source sentence, at the cost of sometimes not being able to find the exact translation.

4.4 Conclusions

We have presented a straightforward system combination method using several well-known feature functions for rescoring the 1-best output of the phrase- and Ngram-based SMT systems. The TALPcom is the combination of the TALPphr and the TALPtup, using several n-gram language models, a word bonus and the IBM Model 1 for the whole sentence. The combination seems to obtain clear improvements in BLEU score, but not in NIST, since the features that operate in the combination generally benefit shorter outputs.

We have reported a structural comparison between the phrase- and Ngram-based system. On the one hand, the Ngram-based system outperforms the phrase-based in terms of search time efficiency by avoiding the overpopulation problem presented in the phrase-based approach. On the other hand, the phrase-based system shows a better performance when decoding under a highly constrained search.

We have carried out a detailed error analysis in order to better determine the differences in performance of both systems. The Ngram based system produced more accurate translations, but also a larger amount of extra (incorrect) words when compared to the phrase-based translation system.

Finally, we have presented another system combination method which consists of concatenating a list of the respective system outputs and rescoring them using the opposite system as a feature function, i.e. the Ngram-based system is used for the phrase-based system and vice-versa. For both systems, including the probability given by the opposite system as a rescoring feature function leads to an improvement of BLEU score.

RELATED PUBLICATIONS

- Crego,J.M., Costa-jussà,M.R., Mariño,J.B. and Fonollosa, J.A.R.
  Ngram-based versus Phrase-based Statistical Machine Translation
  International Workshop on Spoken Language Translation (IWSLT) pp 177-184, Pittsburgh, 2005.

- Costa-jussà, M.R., Crego, J.M., de Gispert, A., Lambert, P., Khalilov, M., Fonollosa, J.A.R., Mariño,J-B,and Banchs, R.
  TALP Phrase-based Statistical Machine Translation and TALP system combination the IWSLT 2006
  International Workshop on Spoken Language Translation (IWSLT) pp 123-129, Kyoto, November 2006.

- Costa-jussà, M.R., Crego, J.M., Vilar, D., Fonollosa, J.A.R., Mariño, J.B. and Ney, H..
  Analysis and System Combination of Phrase- and Ngram-based Statistical Machine Translation Systems
  HLT-NAACL Conference pp 37-140, Rochester, May 2007

5

State-of-the-art Reordering Approaches

This chapter describes several state-of-the-art reordering techniques employed in SMT systems. Reordering is understood as the word order redistribution of the translated words as shown in Figure 5.1. In initial SMT systems, this different order is only modeled within the limits of translation units.

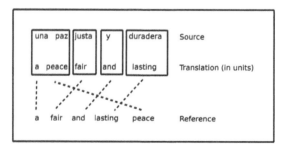

Figure 5.1: *Source, translation (in units) and reference example. Translation and reference differ in word order.*

Relying only in the reordering provided by translation units may not be good enough in most language pairs, which might require longer reorderings (as shown in Figure 5.1). Therefore, additional techniques may be deployed to face the reordering challenge. That is why many extended approaches (Chang and Toutanova, 2007) propose to face statistical machine translation as a concatenation of two sub-tasks: predicting the collection of words in a translation and deciding the order of the predicted words.

This chapter is organized as follows. Section 5.1 focuses on the description of several reordering approaches and Section 5.2 sets the guidelines for the novel reordering algorithms that are described in Chapters 6 and 7.

5.1   Description

As previously introduced, reordering between two languages is a widely studied challenge in MT since different languages have different word order requirements.

In recent years, several alternatives to tackle the word ordering in translation have been proposed. These alternatives may be classified into the following groups:

- **Straight heuristic reordering search constraints**, which are founded on the application of distance-based restrictions to the search space.

- **Source reordering approaches**, where reordering rules are defined in the source language. The idea is to reorder the source language in a way that better matches the target language.

- **Reordering in rescoring**, typically the rescoring methods have generally provided small accuracy gains given the restriction of being applied to an N-best list.

- **Reordering based on syntax structures**, which is not carried out using standard phrases. In fact, it solves translation following dependency trees.

In the next lines, several reordering approaches are described following the above classification criteria. It is worth noticing that this classification is clearly subjective because there are no clear boundaries among categories. For the sake of simplicity a reordering approach is only included in one category.

5.1.1 Straight Heuristic Reordering Search Constraints

Many systems use very simple models to reorder translation units. The straight reordering approach is to introduce in the search space multiple permutations of the input sentence, aiming at acquiring the right word order of the resulting target sentence. However, systems are forced to restrict their distortion abilities because of the high cost in decoding time that permutations imply. In (Knight, 1999a), decoding with arbitrary word reorderings is shown to be NP-hard and, at the unit level, reordering is still computationally expensive.

Therefore, different distance-based constraints were commonly used to make the search feasible. The use of these constraints generates simple reordering models which imply a necessary balance between translation accuracy and efficiency. One simple model is a 'weak' distance-based distortion model that was initially used to penalize the longest reorderings, only allowed if sufficiently promoted by the rest of models: distortion (Och and Ney, 2004; Koehn et al., 2003); IBM(Berger et al., 1996; Tillmann and Ney, 2003); ITG (Wu, 1996); local (Kanthak et al., 2005) or maxjumps (Crego et al., 2005b).

IBM constraints are allowed to deviate from the monotonic order by postponing translations up to a limited number of words, *i.e.* at each state, translations can be performed on the first l word positions not yet covered. At each state, the covered words are shown in the form of a bit vector. Figure 5.2 shows the permutations graph computed for a monotonic (top) and a reordered (bottom) search of an input sentence of $J = 4$ words. The reordered graph shows the valid permutations computed following IBM constraints, defined in (Zens and Ney, 2003), for a value of $l = 2$.

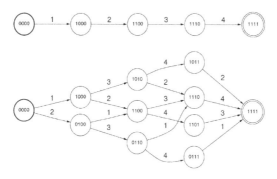

Figure 5.2: *Permutations graph of a monotonic (top) and reordered (bottom)* search.

Another simple reordering model is the flat model: ITG (Wu, 1996), which is not content dependent either. Flat model assigns constant probabilities for monotone and non-monotone order. The two probabilities can be set to prefer monotone or non-monotone orientations, depending on the language pairs. (Zens and Ney, 2003) shows a comparison between IBM and ITG constraints. (Xiong et al., 2006) proposes a novel solution for unit reordering. Under the ITG constraint they need to consider just two kinds of reorderings, straight and inverted between two consecutive blocks. In this way, reordering can be modeled as a problem of classification with only two labels. The main drawback is that not all reorderings can be captured (Wellington et al., 2006).

The maxjumps approach which was used in Chapter 3 limits the number of reorderings for a search path (whole translation) to a given number using two strategies:

- A distortion limit (m). A source word (phrase or tuple) is only allowed to be reordered if it does not exceed a distortion limit, measured in words.

- A reorderings limit (j). Any translation path is only allowed to perform j reordering jumps.

In particular, when using this approach in an Ngram-based system, the unfolding technique (see § 2.3.2) is computed to extract translation units.

In view of content-independence of the distortion and flat reordering models, several researchers (Tillmann, 2004; Koehn et al., 2005) proposed a more powerful model called lexicalized reordering model that is phrase dependent. Lexicalized reordering model learns local orientations (monotone or non-monotone) with probabilities for each bilingual phrase from training data. During decoding the model attempts to finding a Viterbi local orientation sequence. Performance gains have been reported for systems with lexicalized reordering model. However, since reorderings are related to concrete phrases, researchers have to design their systems carefully in order not to cause other problems, e.g. the data sparseness problem. This lexicalized reordering approach is implemented in the open source Moses toolkit [1].

Additionally to the reordering constraints, a reordering approach is defined by a distortion model. The distortion models allow to assign a score to each reordering constraint or permutation explored by the search. A standard distance-based distortion model is defined as follows (Crego et al., 2005b):

---

[1] http://www.statmt.org/moses/

$$P(u_1^K) = exp(- \sum_{k=1}^{K} d_k)$$

where d_k is the distance between the first word of the kth translation unit (u), and the last word+1 of the $(k - 1)^{th}$ translation unit. To sum up, the main idea of distance-based reorderings is that if N words are skipped, a penalty of N will be paid regardless of which words are reordered. This model takes the risk of penalizing long distance jumps which are common between two languages with very different orders.

5.1.2 Source Reordering Approaches

Similar to the previous heuristic search constraints, the reordering alternative detailed in this section aims at applying a set of permutations to the words of the input sentence to help the system build the translation hypothesis in the right word order. The main difference here is that these approaches attempt to specifically learn the source reordering that better matches the target language. The reordering rules and/or constraints in the source language may be defined following different criteria such as:

- DETERMINISTIC REORDERING. Word order harmonization was first proposed in (Nießen and Ney, 2001), where morpho-syntactic information was used to account for the reorderings needed between German and English. In this work reordering was done by prepending German verb prefixes and by treating interrogative sentences using syntactic information. In (Lee et al., 2006) and (Popovic and Ney, 2006), the source corpus is reordered following a set of rules. In the former, the rules are lexicalized, in the latter, these rules have been automatically learned using lexical and/or morphological information, i.e. *Part of Speech* (POS). The decoder search is monotonic. Figure 5.3 (top) shows how reordering and decoding are decoupled under this approach in two main blocks. One of the main drawbacks of this approach is that it takes reordering decisions in a preprocessing step, thus discarding much of the information available in the global search that could play an important role if it were taken into account. Thus far, the reordering challenge is only tackled in preprocessing, so the errors introduced in this step may remain in the final translation output.

- CLAUSE RESTRUCTURING. Here, the methods use syntactic information to reorder source words in SMT as a preprocessing step. (Xia and McCord, 2004) proposes a set of automatically learned reordering rules (using morpho-syntactic information in the form of

POS tags) which are then applied to a French-English translation task. In (Collins et al., 2005) a German parse tree is used for moving German verbs towards the beginning of the clause. (Wang et al., 2007) uses manually written Chinese rules for moving Chinese structure towards the English one. (Habash, 2007) employs dependency trees to capture reordering needs of an Arabic-to-English translation system. In these works, the source reordering is complemented with a local reordering in search.

- INPUT REORDERING GRAPH. A natural evolution of the harmonization strategy is shown in Figure 5.3 (bottom). It consists of using a word graph, containing the N-best reordering decisions, instead of the single-best used in the above strategy. The reordering problem is equally approached by alleviating the difficulty of needing highly accurate reordering decisions in preprocessing. The final decision is delayed, to be subsequently in the global search, where all the information is then available. Inspired by (Knight and Al-Onaizan, 1998), they permute the source sentence to provide a source input graph that extends the search graph. In (Kanthak et al., 2005), they train the system using a monotonized source corpora and they translate the test set allowing source reorderings which are limited by constraints such as IBM or ITG as used in §5.1.1. Similarly in (Crego, 2008; Zhang et al., 2007), reordering is addressed through a source input graph. In this case, the reordering hypotheses are defined from a set of linguistically motivated rules (either using *Part of Speech*; chunks; or parse trees).

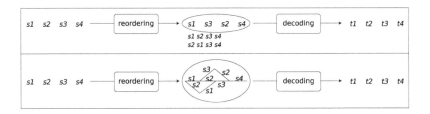

Figure 5.3: *Source reordering approaches.*

5.1.3   Reordering in Rescoring

Rescoring techniques have also been proposed as a method for using syntactic information to identify translation hypotheses expressed in the right target word order (Koehn and Knight, 2003a; Och et al., 2004; Shen et al., 2004). In these approaches, a baseline system is used to generate N-best translation hypotheses. Syntactic features are then used in a second model that re-ranks the N-best lists, in an attempt to improve over the baseline approach. (Koehn and Knight, 2003a) apply a re-ranking approach to the sub-task of noun-phrase translation.

(Hassan et al., 2006) introduces super-tag information or *'almost parsing'* (Bangalore and Joshi, 1999) into a standard phrase-based SMT system in the re-ranking process. It is shown how syntactic constraints can improve translation quality for an Arabic-to-English translation task. Later, in (Hassan et al., 2007), the same researchers introduce the super-tag information into the overall search in the form of an additional log-linearly combined model.

Another kind of approach, which may be included in § 5.1.4 is to use syntactic information in rescoring methods. (Koehn and Knight, 2003b) apply a reranking approach to the sub-task of noun-phrase translation. (Shen et al., 2004) describes the use of syntactic features in reranking the output of a full translation system, but the syntactic features give very small gains.

5.1.4   Reordering based on Syntax Structures

In spite of the great success of the phrase-based systems, a key limitation of these systems is that they make little or no direct use of syntactic information. However, it appears likely that syntactic information can be of great help in order to accurately model many systematic differences (Bonnie, 1994) between the word order of different languages. Ideally, a broadly covered and linguistically well-motivated statistical MT system can be constructed by combining the natural language syntax and machine learning methods.

In recent years, syntax-based statistical machine translation has begun to emerge, aiming at applying statistical models to structured data. Advances in natural language parsing, especially the broad-coverage parsers trained from tree banks, for example (Collins, 1999), have made possible the utilization of structural analysis of different languages.

Several methods have been proposed that use syntactic information to handle reordering. In this context, a syntax-directed translator consists of two components, a source language parser and a recursive converter, which is usually modeled as a top-down tree-to-string transducer. All these approaches, though different in formalism, model the two languages using tree-based transduction rules or a synchronous grammar, possibly probabilistic. Machine translation is done either as a stochastic tree-to-tree transduction or synchronous parsing process. A further decomposition of these systems can be done by looking at the kind of information they employ.

In (Quirk et al., 2005; Langlais and Gotti, 2006), a dependency tree-based reordering model is inferred from aligned string-tree pairs. Parsing is performed on the source language and a corresponding dependency grammar is inferred on the aligned target side.

Others use constituent trees (Chiang, 2005; Watanabe et al., 2006) in which context-free rules are inferred from string-to-string pairs (notice: no parsing is required). In this approach, phrases are reorganized into hierarchical by reducing subphrases to variables. This template-based scheme not only captures the reorderings of phrases, but also integrates some phrasal generalizations into the global model.

Some methods are linguistically syntax-based (Yamada and Knight, 2002; Wu, 1997; Marcu et al., 2006). One approach makes use of bitext grammars to parse both the source and target languages. Another approach makes use of syntactic information only in the target language. In the syntax model in (Galley and Hopkins, 2004), syntactic translation rules are inferred from aligned tree-string pairs and parse trees are computed on the target language.

Note that these models have structures and parametrization that are radically different from those of phrase-based models for SMT.

Therefore, syntax-based decoders have emerged that aim at dealing with pairs of languages with very different syntactic structures for which the word context introduced in phrase-based decoders is not sufficient to cope with long reorderings. They have gained much popularity because of the significant improvements made by exploiting the power of synchronous rewriting systems.

Syntax-directed systems have been typically attacked with the argument of showing a main weakness in their poor efficiency results. However, this argument has been recently overridden through the development of new decoders, which show significant improvements when handling syntactically divergent language pairs under large-scale data translation tasks. An example of such a system can be found in (Marcu et al., 2006), which has obtained state-of-the-art results in Arabic-to-English and Chinese-to-English large-sized data tasks.

5.2 Our Contribution

Chapters 6 and 7 propose two different reordering techniques that constitute the core work of this PhD.

Machine translation is the automation of a human task. The big question here is how the human translation process work. One extended theory is that: *The translator will undo the syntactic structure of the source text and then formulate the corresponding message in the target language* [2] As a parallel to this, the source reordering approaches presented above try to undo the syntactic structure of the source text in order to better match the target text. In this PhD we follow this research line.

Our first approach could be included in the category of source reordering approaches and deterministic reordering rules, where the main novelty is that the rules are learned statistically from the training set and new rules can be inferred without requiring additional knowledge.

Our second approach: Statistical Machine Reordering (SMR) is an Ngram-based reordering technique. It confronts the reordering challenge using the powerful statistical machine translation techniques. It could be included in the category of source reordering approaches and input reordering graph. Notably, the novel SMR technique offers the following main advantages:

1. Reorderings are learned from the aligned parallel corpus. While most methods and reordering hypotheses are confronted with a data sparseness problem, the SMR approach smooths these by taking advantage of the extensively investigated area of language modeling.

2. The SMT translation is nearly as efficient as a monotonic translation because, through the use of statistical criteria the input reordering graph can be highly pruned without affecting the translation quality.

3. Each reordering hypothesis has a probability that can be included in the SMT log-linear framework.

4. The SMR approach offers some versatility. Depending on the pair of languages, the reordering hypothesis may be better captured by using a particular type of word class, i.e. statistical or morphological.

Further advantages and limitations of both approaches are fully presented and discussed in Chapters 6 and 7.

---

[2]http://accurapid.com/journal/05theory.htm

RELATED PUBLICATIONS

- *State-of-the-art Word Reordering Approaches in Statistical Machine Translation*
  Costa-jussà, M.R. and Fonollosa, J.A.R.
  IEICE Transactions on Information and Systems vol 92, num 11, pp 2179-2185, 2009

6

Recursive Alignment Block Classification Technique

Statistical Machine Translation (SMT) is based on alignment models that learn the word correspondences between the source and target languages from bilingual corpora. These models are assumed to be capable of learning reorderings of the sequences of words. However, the difference in word order between two languages is one of the most important sources of errors in SMT. As reported in the introduction, the main goal of this thesis is to address this typology of errors. This chapter presents a first attempt to solve reordering problems by using inductive learning. We propose a Recursive Alignment Block Classification technique (RABCA). This technique should be able to cope with swapping the examples seen during training; it should infer properties that might allow for reordering in pairs of blocks (i.e., sequences of words) not seen together during training. Finally it should be robust with respect to training errors and ambiguities.

This chapter is organized as follows. Section 6.1 presents the motivation for the reordering technique. Section 6.2 reports its introduction into a standard SMT system. Section 6.3 describes several experiments that assess the quality of the approach. Section 6.4 summarizes the chapter.

6.1 Motivation

SMT systems are trained by using a bilingual corpus composed of bilingual parallel sentences. Each set of bilingual parallel sentences is composed of a source and a target sentence, and each is aligned at the word level as previously reported in Chapter 2.2. Then, translation units are extracted; and the statistical models are computed.

It is generally known that both the word alignment and the translation perform better when the task is monotonic, i.e., the source and the target have similar word ordering. One way to transform a non-monotonic task into a more monotonic one is by reordering words in the source sentence following the order of words in the target sentence. Hereafter, this process will be referred to as monotonization. For instance in a Spanish to English translation, given the bilingual sentence: *El **discurso político** fue largo # The **political speech** was long*, monotonizing the Spanish sentence would lead to the reordered Spanish sentence: *El **político discurso** fue largo*. As a consequence, the word alignment and translation units are now monotonic: *El#The político#political discurso#speech fue#was largo#long*.

The presented approach reorders the source corpora so that they better match the order of the target corpora. This reordering is done in the source sentence by swapping the Alignment Blocks, i.e., pairs of consecutive sequences of words whose translations appear swapped in the target string. Reordering based on swapping sequences of words covers most cases as shown in (Tillmann and Zhang, 2005). These Alignment Blocks are first learned from the word alignment, which is parsed to detect them. Figure 6.1 presents an example. At this point, a preliminary list of alignment blocks is available. Second, in order to be capable of reordering pairs of blocks that were not seen in the training step, the Alignment Blocks are classified into groups. This classification follows a recursively co-occurrence criterion. Finally, all combinations of blocks belonging to the same group are considered candidates to be swapped, that is, we infer new Alignment Blocks by allowing internal swaps within a group.

6.2 Reordering Description

6.2.1 Enhancing an SMT System with the RABCA Technique

Once the RABCA technique is learned (see §6.2.4), we introduce it in an SMT system as follows:

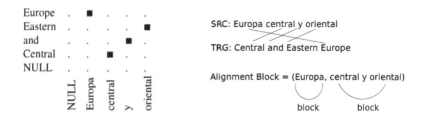

Figure 6.1: *Example of an Alignment Block.*

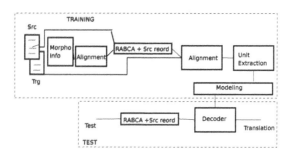

Figure 6.2: *Description system.*

1. Given the candidates to be swapped (Alignment Blocks proposed by the RABCA technique), we reorder the source corpora, including training, development and test sets.

2. Given the reordered source training corpus, we realign it with the original target training corpus. Notice that we compute the word alignment twice: once before extracting the *LAB* and once after the corpus is monotonized to build the SMT system.

3. We build the phrase- and/or Ngram-based systems as shown in Chapters §2.3.1 and §2.3.2, respectively, using the new monotonized corpora.

Figure 6.2 describes the application of the RABCA technique on an SMT system.

6.2.2 Main Steps to Train the RABCA Technique

Note that our reordering technique should be able to cope with swapping the examples seen during training (step 1 below); it should be robust with respect to the noisy training errors and ambiguities (step 2 below); and finally it should infer properties that might allow reordering in pairs of blocks not seen during training (step 3 below). In order to fulfill these expectations, we present the following steps.

1. Given a word alignment, we extract a List of Alignment Blocks (LAB). See Figure 6.1.

2. Given a LAB we obtain a filtered LAB, hereafter referred to as LAB_{filt}, which filters the ambiguous Alignment Blocks (i.e., either misaligned or inherently ambiguous), see §6.2.3.

3. Given the LAB, we apply the Recursive Classification Algorithm. See §6.2.4, to infer new Alignment Blocks.

6.2.3 List of Alignment Blocks (LAB)

As stated before, given a word alignment, we extract a LAB. See Figure 6.1. Given the LAB, we filter the ambiguous Alignment Blocks (i.e., either misaligned or inherently ambiguous). We filter the possible bad alignments or ambiguities using the following criteria:

- Alignment Blocks appearing less than N_{min} times are discarded.

- Alignment Blocks with a swapping probability (P_{swap}) less than a certain threshold are also discarded. We define the swapping probability as the ratio between the number of times that two blocks are swapped and the total number of times that the same two blocks appear consecutively.

6.2.4 Recursive Classification Algorithm

The objective of this algorithm is to extend the list of Alignment Blocks (learned directly from the word alignment and filtered) by following a recursive co-occurrence criteria. By following this criteria we will classify the Alignment Blocks into groups. Then, we infer new Alignment Blocks by allowing internal swaps a group.

We will define the filtered LAB as LAB_{filt}, which will be a subset of LAB and consists of m Alignment Blocks $\{(\alpha_1, \beta_1), (\alpha_2, \beta_2), \ldots, (\alpha_m, \beta_m)\}$. From the LAB_{filt}, we create the sets $A = \{\alpha_1, \alpha_2, \ldots, \alpha_m\}$ and $B = \{\beta_1, \beta_2, \ldots, \beta_m\}$ and the groups $G_1 \ldots G_n \ldots G_N$. A given group G_n is created by recursively following a co-occurrence criterion and has the form $G_n = \{(\alpha_1, \beta_1), \ldots (\alpha_p, \beta_p)\}$ where p is the cardinality of G_n; within each group G_n we also create the sets $A_n = \{\alpha_1, \ldots \alpha_p\}$ and $B_n = \{\beta_1, \ldots \beta_p\}$. From each group G_n we build a generalization group (Gg_n, where $n = 1, \ldots N$) defined as the Cartesian product between the subsets $A_n \in A$ and $B_n \in B$, i.e., $Gg_n = A_n \times B_n$ which will also allow the algorithm to reorder unseen cases such as (α_r, β_s) with $\alpha_r \in A_n$, $\beta_s \in B_n$ and $(\alpha_r, \beta_s) \notin LAB_{filt}$. We can deal with possible inconsistencies, by increasing the filtering threshold, and thereby limiting the number of allowed unseen pairs, and also by processing with morphological information.

Here, we infer the generalization groups Gg_n from the LAB_{filt}. Given the LAB_{filt}, the generalization groups Gg_n are built as follows:

1. Initialization: set $n \leftarrow 1$ and $LAB'_{filt} \leftarrow LAB_{filt}$.

2. Main part: while LAB'_{filt} is not empty do

   - $G_n = \{(\alpha_k, \beta_k)\}$ where (α_k, β_k) is any element of LAB'_{filt}
   - Recursively, move elements (α_i, β_i) from LAB'_{filt} to G_n if there is an element $(\alpha_j, \beta_j) \in G_n$ such that $\alpha_i = \alpha_j$ or $\beta_i = \beta_j$
   - Increase n (i.e., $n \leftarrow n + 1$)

3. Ending: For each G_n, construct the two sets A_n and B_n that consists of the first and second element of the pairs in G_n, respectively. Then the Cartesian product of A_n and B_n is assigned to Gg_n, i.e., $Gg_n \leftarrow A_n \times B_n$.

The next subsection details the present algorithm with an example.

6.2.5   Recursive Classification Algorithm Example

- Suppose a List of Alignment Blocks filtered (LAB_{filt}) as follows.

abogado europeo	($\Rightarrow$ European lawyer)
abogado americano	($\Rightarrow$ American lawyer)
apoyar totalmente	($\Rightarrow$ fully support)
parlamento australiano	($\Rightarrow$ Australian Parliament)
parlamento europeo	($\Rightarrow$ European Parliament)
oponer totalmente	($\Rightarrow$ completely oppose)
traductor americano	($\Rightarrow$ American translator)

- Classify each Alignment Block into a group, using a co-occurrence criterion.

1. Start a loop. Take an unclassified Alignment Block from the LAB_{filt} and create a group.

abogado europeo
abogado americano
apoyar totalmente
parlamento australiano
parlamento europeo
oponer totalmente
traductor americano

Group 1	
A_1	B_1
abogado	europeo

2. Add all the Alignment Blocks to the group that are partially equal (side A or B) with the ones already in the new group.

abogado europeo
abogado americano
apoyar totalmente
parlamento australiano
parlamento europeo
oponer totalmente
traductor americano

Group 1	
A_1	B_1
abogado	europeo
abogado	americano
parlamento	europeo

3. Reach the end of the LAB_{filt}. End of the loop.

   (a) If one or more Alignment Blocks have been classified during the loop, go to step 2.

			Group 1	
abogado europeo			A_1	B_1
abogado americano			abogado	europeo
apoyar totalmente			abogado	americano
parlamento australiano			parlamento	europeo
parlamento europeo			parlamento	australiano
oponer totalmente			traductor	americano
traductor americano				

(b) If no Alignment Block has been classified during the loop and there are still unclassified Alignment Blocks in the LAB_{filt}, go to step 1.

	Group 1		Group 2	
abogado europeo	A_1	B_1	A_2	B_2
abogado americano	abogado	europeo	apoyar	totalmente
apoyar totalmente	abogado	americano	oponer	totalmente
parlamento australiano	parlamento	europeo		
parlamento europeo	traductor	americano		
oponer totalmente	parlamento	australiano		
traductor americano				

(c) If no Alignment Block has been classified during the loop and there are no unclassified Alignment Blocks in the LAB_{filt}, stop.

- Once all the Alignment Blocks in the LAB_{filt} are classified, the final groups are:

Group 1		Group 2	
A_1	B_1	A_2	B_2
abogado	europeo	apoyar	totalmente
parlamento	americano	oponer	
traductor	australiano		

- Within the same group, we allow new internal combination in order to generalize the reordering to unseen sequences of words. These are the generalization groups. We can see the generalizations to unseen reorderings in Group 1: *abogado australiano*, *parlamento americano*, *traductor australiano* and *traductor europeo*.

- The groups seem to learn morphological information:

  1. Group 1 = Noun + Adjective
  2. Group 2 = Verb + Adverb

6.2.6   Using Extra Morphological Information

The Alignment Block Classification may be applied to a lemmatized corpus. In this case, the final Gg_n could deal with gender or number disagreements between the elements of each block. For instance the pair (*conferencia, parlamentario*), which does not have gender agreement, would be a correct and useful generalization if applied to a lemmatized source. Therefore the generalization would not be influenced by the particular distribution of word inflexions in the training database.

Additionally, a tag on each word could be used to filter generalizations. In experiments in this chapter(§6.3), we will describe an example.

6.3   Experimental Work

This section details the experiments carried out to assess the translation accuracy of the proposed reordering algorithm. Experiments have been carried out by considering the European Plenary Parliamentary Speeches (EPPS) Spanish to English task. This translation task conveys only local reorderings. Full details of the corpora employed for experiments are presented in A.3.

6.3.1   Baseline SMT Systems.

	Phrase-based system
Phrase-length	10
max target per source	20
Translation model	Conditional and posterior probability
Features	s2t and t2s lexicon models, word and phrase bonus, target language model POS target language model
Beam	50
Search	Monotonic

	Ngram-based system
	no limit
Tuple-length	no limit
solving NULL	IBM-1 approach
max target per source	20
Translation language model	4gram kneser-ney interpolation
Features	s2t and t2s lexicon models, word bonus, target language model POS target language model
Beam	50
Search	Monotonic

Table 6.1: *Phrase (left) and Ngram (right) default parameters.*

As stated before, phrase- and Ngram-based systems were used as baseline systems. Given a parallel corpus, standard tools were used for tokenizing and filtering. The English side of the training corpus was POS tagged using the freely available TnT tagger (Brants, 2000), and the Spanish side was POS tagged and lemmatized using the freely available FreeLing[1]

---

[1]http://www.lsi.upc.edu/~nlp/freeling/

tool (Carreras et al., 2004).

After preprocessing the training corpora, word-to-word alignments were performed in both alignment directions using Giza++ (Och and Ney, 2003), and the union set of both alignment directions was computed. Phrase/Tuple sets for each translation direction were extracted from the union set of alignments. The resulting phrase/tuple vocabularies were pruned by considering the N best translations for each tuple source-side ($N = 20$ for the Spanish-to-English direction) in terms of occurrences.

The SRI[2] language modeling toolkit (Stolcke, 2002) was used to compute all N-gram language models (including our special translation model). Kneser-Ney smoothing (Kneser and Ney, 1995) and interpolation of higher and lower N-grams were always used for estimating the translation N-gram language models.

Once the models were computed, optimal log-linear coefficients were estimated for each system configuration using an in-house implementation of the widely used downhill Simplex method (see single loop optimization in §2.5). The BLEU score was used as the objective function.

The decoder was always set to perform histogram pruning, keeping the best $b = 50$ hypotheses (during the optimization work, histogram pruning was set to keep the best $b = 10$ hypotheses).

General parameters of both baseline systems are show in Table 6.1.

6.3.2 RABCA Experiments.

We have looked for the most common reordering patterns found in our task. We have a reference corpus that consists of 500 manually aligned bilingual sentences (Lambert et al., 2006). Given the word alignment reference, we can extract the reordering patterns. Most common reordering patterns have been described as: $(x_1, y_1)(x_2, y_2)...(x_N, y_N)$ where each (x_i, y_i) describes a link between the position x_i and y_i, in the original and the reordered source sentence composed of the source words appearing in the monotonization of the alignment. This means that the cross $(0,1)(1,0)$ would reflect: $a_n\ b_n$ to $b_n\ a_n$, where a_n (b_n) is only one word. Table 6.2 presents the most frequent reordering patterns when aligning from Spanish to English with the EPPS task. Most of them can be resolved by swapping sequences of words. In particular, in next experiment we deal with the most frequent reordering pattern: $(0,1)(1,0)$.

---

[2]http://www.speech.sri.com/projects/srilm/

Reord. pattern SPA->ENG	Counts	%
(0,1)(1,0)	392	38.4%
(0,2)(1,0)(2,1)	113	11%
(0,1)(1,2)(2,0)	112	11%
(0,2)(1,3)(2,0)(3,1)	38	3.7%
(0,3)(1,0)(2,1)(3,2)	37	3.7%
(0,2)(1,1)(2,0)	25	2.4%
(0,3)(1,4)(2,0)(3,1)(4,2)	16	1.6%
Most freq. patterns	733	71.8%

Table 6.2: *Reordering patterns for Es2En reference alignment of 500 sentences.*

As stated above, morphological information is already available for this language pair, and, we took advantage of it. We lemmatized the parallel corpus and performed the first word alignment from Figure 6.2 using the lemmatized corpus. We extracted the *LAB*, limiting the complete Alignment Block length to two words and considering only the most frequent pattern of our task. Alternatively, we experiment to remove those pairs of blocks that are not constituted by *Noun* plus *Adjective*. As shown later, this rule achieves good results. In general, *Noun* plus *Adjective* in Spanish becomes *Adjective* plus *Noun* in English. This rule should not be applied independently as there are a few common exceptions, such as *gran hombre (big man)*.

6.3.2.1   Tuning RABCA Parameters: LAB Filtering Parameters N_{min} and P_{swap}.

The amount of admissible Alignment Blocks in the LAB_{filt}, is a function of the parameters N_{min} and P_{swap} (see subsection §6.2.4). We determine these parameters from a subset of the corpus as follows. We remove the 500 manually aligned sentences from the training corpus. We train the RABCA technique and swap the reference source set. Given a swapping of two words, it can be a *Success* (S) if the reference alignment is swapped, or a *Failure* (F) if the reference alignment is not swapped. Combining these two sources of information, we use the simplex algorithm to minimize the following:

$$Q = N_S - N_F \tag{6.1}$$

We chose the cost function Q as a coherent criterion to optimize the number of successes (N_S) and minimize the number of failures (N_F). The cost function Q has two quantified variables as an argument, and its output is a difference between two integers. Note that the underly-

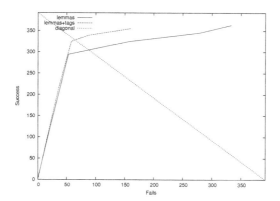

Figure 6.3: *Relation between successes and fails (with different parameters -N_{min}, P_{swap}- for the RACBA: (1) using lemmas and (2) using lemmas plus tags), for the manually aligned reference corpus.*

ing problem is a multi-objective optimization, which we transformed to a simple optimization problem by giving equal importance to the two objectives; i.e., $Successes$ and $-Failures$. For this kind of problems, direct search techniques such as the simplex algorithm are adequate.

Figure 6.3 shows the relation between the two objectives, which gives a curve similar to the ROC curve used in detection theory. An increase in the success rate increases the failure rate in ROC, therefore there is a trade-off between two objectives. The solution that we have selected is the intersection of the diagonal with the curve, which corresponds to a trade-off that gives the same weight to both objectives. The maximum Q corresponds to the curve of lemmas reordering plus tags. Figure 6.3 shows that the *LAB* filtering parameters that minimize Q are $N_{min} = 5$ and $P_{swap} = 0.53$.

Given the optimum values of N_{min} and P_{swap}, we used the Recursive Classification Algorithm to generate the Generalization groups. We have also studied the number of good generalizations, i.e., the pairs of words that have been swapped correctly and that were not seen swapped during training: almost half of the $Successes$ are generalizations.

6.3.2.2   Enhancing Phrase- and Ngram-based SMT Systems with the RABCA Technique.

In order to evaluate the method we developed, we introduced the RABCA reordering in the SMT baseline systems as shown in Figure 6.2. We followed these steps:

1. Reorder all the Spanish corpora (training, development and testing) using the RABCA approach.

2. Use GIZA++ to realign the training parallel corpus.

3. Apply the baseline SMT systems using the new alignment.

4. Optimize with the reordered development and translate the reordered test.

Experiment	AER
Alignment	20.64
Alignment + RABCA	**19.36**

Table 6.3: *AER results in the EPPS Es2En task.*

Next, we compared the word alignment quality of the baseline and the enhanced system. Evaluation was conducted with the 500 hundred manually aligned sentences cited above. Table 6.3 shows the improvement in alignment error rate (AER).

System	Configuration	mWER	BLEU
PB	Baseline	34.44	55.23
PB	Baseline + RABCA	**33.75**	**56.32**
NB	Baseline	34.46	55.24
NB	Baseline + RABCA	**33.68**	**56.26**

Table 6.4: *Results in the EPPS Es2En task for the phrase- and Ngram-based system.*

Table 6.4 shows the translation results. Both measures, mWER and BLEU, improve around one point by using *RABCA*. The task is relatively monotonic. Therefore, reordering is mainly local. However, both mWER and BLEU improve significantly. When comparing the RABCA performance in both systems, we see that the translation quality is similarly improved in both cases.

6.4   Conclusions and Further Work

This chapter introduced a local reordering approach based on monotonization of the source corpus, which benefits alignment and translation quality.

The proposed reordering improves the alignment itself because it monotonizes the bilingual text, so the alignment becomes more robust. The reordering process depends on word alignment quality. However, the process of filtering the List of Alignment Blocks allows us to reduce noisy alignment and training ambiguities.

RABCA improves the translation quality by providing more accurate local reorderings than the ones provided by only the translation units (either phrases or tuples).

Moreover, RABCA through the Recursive Classification algorithm is able to generalize to unseen reorderings.

Both measures, mWER and BLEU, improve significantly. This task is relatively monotonic. Nevertheless, adding local reordering to the SMT systems leads to a better translation performance.

Further improvements could be expected with longer blocks and by experimenting with language pairs that are less monotonic. We leave for further research the following points:

1. Do experiments with longer blocks.

2. Evaluate the system with tasks with greater reordering difficulties such as the pair Arabic/English.

3. Deal with the inference of rules based on a wider context, in the case that morphological information is available.

One main drawback of this approach is that the reordering is given to the decoder that makes a monotonic translation.

RELATED PUBLICATIONS

- *Phrase and Ngram-based Statistical Machine Translation System Combination*
  Costa-jussà, M.R. and Fonollosa, J.A.R.
  Applied Artificial Intelligence: An International Journal vol 23, num 7, pp 694-711, 2009

- *Using Reordering in Statistical Machine Translation based on Alignment Block Classification*
  Costa-jussà, M.R. and Fonollosa, J.A.R. and Monte, E.
  LREC Conference, Marraqueix, May 2008.

- *TALP Phrase-based statistical translation system for European language pairs*
  Costa-jussà, M.R., Crego, J.M., de Gispert, A., Lambert, P., Khalilov, M., Mariño, J.B., Fonollosa, J.A.R., and Banchs, R.
  HLT/NAACL 2006 Workshop on Statistical Machine Translation (WMT'06), pp. 142-145. New York City, June 2006.

- *UPC's Bilingual N-gram Translation System*
  Mariño,J.B., Banchs,R., Crego, J.M., de Gispert, A., Lambert, P., Fonollosa, J.A.R, Costa-jussà, M.R. and Khalilov, M.
  TC-Star Workshop on Speech-to-Speech Translation TCSTAR'06 pp 43-48 Barcelona, June 2006.

7

Statistical Machine Reordering

This chapter describes a novel approach to solving the reordering challenge. Following a monotonization technique (see Chapter 6), the main idea of the present approach is to convert the entire source corpus into an intermediate representation, in which source-language words are presented in an order that more closely matches that of the target language.

One major factor in this approach is the use of powerful SMT techniques to generate this intermediate representation. Therefore, this approach is called Statistical Machine Reordering (SMR). In order to achieve the desired generalization, the developed approach makes use of word classes. Because it is necessary to couple the SMR and SMT systems, different coupling strategies and new reordering features are presented and added to the final SMT search.

This chapter is organized as follows. Section 7.1 details the fundamentals of the SMR approach. Section 7.2 describes the type of classes introduced in the SMR system. Section 7.3 presents different ways to couple the reordering and translation systems. Section 7.4 reports on the experiments conducted to asses the accuracy and efficiency of the SMR approach and compares it to the lexicalized reordering approach implemented in Moses. Experiments are automatically evaluated and manually analyzed to report errors. Section 7.5 presents the chapter conclusions.

7.1   Baseline SMR System

This section describes the statistical machine reordering approach. The aim of the SMR technique is to use an SMT system to deal with reordering problems. Therefore, the SMR system can be seen as an SMT system that translates from an original source language (S) into a reordered source language (S'), given a target language (T).

This new approach, which can be modeled with the standard noisy channel, models the probability of a reordered source language sentence $s'^J_1 = s'_1 \ldots s'_J$ given a source language sentence $s^J_1 = s_1 \ldots s_J$ as follows:

$$\tilde{s}'^J_1 = \underset{s'^J_1}{argmax} \; \frac{p(s^J_1|s'^J_1)\, p(s'^J_1)}{p(s^J_1)} \tag{7.1}$$

We can ignore the denominator $p(s^J_1)$ inside the *argmax* since we are choosing the best s'^J_1 sentence for a fixed sentence s^J_1, and hence $p(s^J_1)$ is a constant.

$$\tilde{s}'^J_1 = \underset{s'^J_1}{argmax} \; p(s^J_1|s'^J_1)\, p(s'^J_1) \tag{7.2}$$

Here $p(s'^J_1)$ is the language model of the reordered source language. Then, $p(s^J_1|s'^J_1)$ is the string *Source-to-Reordered Source* ($S2S'$) model, which is the basis of the reordering.

Again, such an approach has been expanded to a more general maximum entropy approach in which a log-linear combination of multiple feature functions is implemented (Och, 2003). The overall architecture of this statistical translation approach is summarized in Figure 7.1.

$$\tilde{s}' = \underset{s'}{argmax} \left\{ \sum_{m=1}^{M} \lambda_m h_m(s', s) \right\} \tag{7.3}$$

7.1.1   SMR Model Training

The SMR system uses a bilingual ($S2S'$) n-gram language model (hereafter, SMR Model) to translate from S to S'. This SMR Model is learned in a similar manner to the bilingual n-gram translation language model. Here, the bilingual units contain reordering information. The last sentence in Figure 7.2 is a SMR bilingual unit and, therefore, is a reordering hypothesis. Note

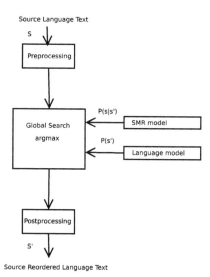

Figure 7.1: *Architecture of the SMR approach based on the log-linear framework approximation*

that a bilingual unit of length 1 does not generate any reordering change. Therefore, the length of the reordering that can be captured depends directly on the length of the SMR bilingual units. Theoretically, the SMR approach manages to deal with local as well as long reorderings because there is no limit on the SMR bilingual units size. In practice, this depends mainly on the ability of the word classes to generalize.

Given the source and target training corpora (parallel at the sentence level), the SMR Model is trained as follows (see Figure 7.1.1):

1. Use the source corpus to train statistical word classes (Och, 1999).

2. Align parallel training sentences at the word level.

3. Extract reordering tuples, see Figure 7.2.

   (a) From the word alignment and the following criteria in § 2.3.2, extract bilingual *S2T* (source-to-target) units while keeping the word alignment information. Figure 7.2 (A) shows an example.

(A) BILINGUAL S2T TUPLE:

better and different structure # estructura mejor y differente # 0-0 0-1 1-2 2-3 3-0

(B) MANY-TO-MANY WORD ALIGNMENT ⟶ MANY-TO-ONE WORD ALIGNMENT:

better and different structure # estructura mejor y differente # 0-1 1-2 2-3 3-0

(C) BILINGUAL S2S' TUPLE:

better and different structure # 1 2 3 0

(D) CLASS REPLACING:

C36 C88 C185 C176 # 1 2 3 0

Figure 7.2: *Example of the extraction of SMR bilingual units. In (a) and (b) '#' divides the fields: source, target and word alignment, which includes the source and final position separated by '-'. In (c) and (d) '#' divides the source and positions of the reordered source.*

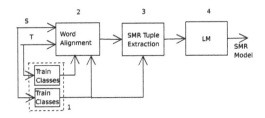

Figure 7.3: *Block diagram of the training process of the SMR model.*

(b) Modify the many-to-many word alignment to many-to-one. If one source word is aligned to two or more target words, the most probable link given IBM-1 lexical probabilities (P_{ibm1}) is chosen, while the others are omitted. Since P_{ibm1}(better, mejor) is higher than P_{ibm1}(better, and), Figure 7.2 (A) leads to Figure 7.2 (B).

(c) From these *S2T* units, extract *S2S'* units that consist of a source fragment and its reordering. See Figure 7.2 (C).

(d) Eliminate bilingual units whose source fragment consists of the NULL word.

(e) Replace the words of each source fragment with the classes determined in the first step of this enumeration. See Figure 7.2 (D).

4. Compute the SMR Model using standard *n*-gram language modeling techniques, given the *S2S'* sequence.

7.1.1.1   Additional Feature Functions

SMR is built by considering a log-linear framework of probabilistic information. Therefore, feature functions can be added easily.

Following Figure 7.1, the main feature function in an Ngram-based system is:

$$h_{smr}(s'^{J}_{1}, s^{J}_{1}) = log \; p(s'^{J}_{1}|s'^{J}_{1}) \qquad (7.4)$$

As the main model (SMR model) is already a language model, no additional models are required. Yet we may use another feature function to improve the performance:

- Additional monolingual language models which are trained on the monotonized source corpus.

7.1.2   SMR Module: from *Source* to *Reordered Source* Language

The SMR module is in charge of translating the source sentence(S) into a reordered source sentence (S'). Figure 7.1.1 shows the block diagram, and each block works as follows:

1. Class replacement. Use the correspondence of word to word class to substitute each source word by its word class.

2. Decoding. A monotonic decoding using the SMR Model allows to assign reordering tuples to the input sequence.

3. Post Processing. The decoder output is post-processed to build the reordered sentence.

An example of the input and output of each step is shown in Figure 7.4. The SMR module can output either a single sentence or a word graph. The former is a reordered sentence like the one shown in the fourth row of Figure 7.4. This gives a unique reordering option, which is the reordered source sentence (S'), and this leads to a deterministic reordering. On the other hand, the latter contains several possible reorderings coded in a graph (see an example of an SMR output graph in Figure 7.5).

S: *Sólo una sociedad plural , democrática y segura puede garantizar el ejercicio pleno de las libertades .*

CLASS REPLACING: *4 166 197 140 53 71 67 112 159 115 155 134 39 43 127 30*

DECODING: *4# | 166#0 | 197 140 53 71 67 112#1 2 3 4 5 0 | 119#0 | 159#0 | 115#0 | 155#0 | 134#0*

| 39#0 | 43#0 | 127#0 | 30#0

POST PROCESSING (S'): *Sólo una plural , democrática y segura sociedad puede garantizar el ejercicio*

pleno de las libertades.

Figure 7.4: *Example of a source sentence reordering performed by the SMR module. The decoding output is shown in units: '|' marks the unit boundaries and '#' marks the two components of the bilingual units. Note that the reordered sentence follows the order of the reference target sentence: Only a plural , democratic and secure society can guarantee the full exercise of freedoms.*

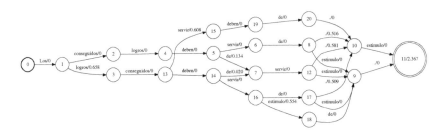

Figure 7.5: *SMR output graph. The source sentence is:* Los logros conseguidos deben servir de estímulo. *The target sentence could be:* The achieved goals should be an encouragement.

7.2 Word Classes

Various clustering techniques have been proposed in natural language processing to solve the challenge presented by sparse data. Statistical machine reordering also faces this kind of challenge. Our aim is to use word classes to allow for more robust reorderings and to be capable of generalizing them, i.e., reorder sequences of words that have not appeared during training.

We mainly propose to use two types of classes: statistical and morphological. Both are learned in an statistical manner under supervised or unsupervised method.

7.2.1 Statistical Classes

Initial SMR studies (Costa-jussà and Fonollosa, 2006; R. Costa-jussà and R. Fonollosa, 2007) considered only statistical classes that have a unique correspondence. We used an efficient method for statistical word clustering that was presented in (Och, 1999) for machine translation [1]. We report a brief explanation in what follows.

7.2.1.1 Statistical Word Clustering Algorithm

The roots of the algorithm can be found in the task of a statistical language model that estimates the probability $Pr(w_1^N)$ of a sequence of words $w_1^N = w_1 \ldots w_N$. A simple approximation of $Pr(w_1^N)$ is to model it as a product of bigram probabilities: $Pr(w_1^N) = \prod_{i=1}^{N} p(w_i|w_{i-1})$. If we want to estimate the bigram probabilities $p(w|w')$ using a realistic natural language corpus we are faced with the problem that most of the bigrams are rarely seen. Using the word class rather than the single word, we avoid the use of the most rarely seen bi-grams to estimate the probabilities. Rewriting the probability using word classes, we obtain the probability model:

$$p = (w_1^N|C) = p(C(w_i)|C(w_{i-1}))p(w_i|C(w_i)) \tag{7.5}$$

The function C maps words to w their classes $C(w)$. This model has two types of probabilities: the transition probability $p(C|C')$ for class C given its predecessor class C' and the membership probability $p(w|C)$ for word w given class C.

The determination of the optimal classes $\hat{C}$ for a given number of classes M is calculated with the following equation:

---

[1] http://www.fjoch.com/mkcls.html

$$\tilde{C} = \underset{c}{argmax}\ p(w_1^N|C) \tag{7.6}$$

An efficient optimization algorithm for this purpose is the exchange algorithm (Och, 1999).

7.2.1.2   Statistical Word Classes Suitability

These statistical classes have the advantage of being appropriate for any monolingual corpus without additional requirements (no extra information).

Classification of unknown words. Training word classes in a limited data set may generate words without classes in the test set. That is, words appearing in the test set and not in the training set will not have an associated class. Unclassified words appearing in the test set can be solved by using a disambiguation technique. Unknown words are assigned to one class by taking into account their context.

Reordering information. These statistical word classes do not contain any reordering information because they are trained with a monolingual set. One way of incorporating information about order might be by using training classes in the reordered training source corpus. In other words, we monotonize the training corpus with the alignment information (i.e., reorder the source corpus in such a way that it matches the target corpus under the alignment links criterion). See Figure 7.6. We train statistical classes with the monotonized source training, which are hereafter referred to as statistical reordered classes.

SRC: *the economical and political interests were discussed*
TRG: *se discutieron los intereses económicos y políticos*
ALIGN: *0-2 1-4 2-5 3-6 4-3 5-1 6-2*
SRC MONOTONIZATION:*were discussed the interests economical and political*

Figure 7.6: *Example of source monotonization using the alignment information.*

7.2.2   Morphological Classes

In some pair of languages reordering may be related to word morphology as stated in Chapter 6. Therefore, using *Part of Speech* (POS) can capture rules. For example: generally the sequence *Noun Adjective* in Spanish becomes *Adjective Noun* in English. Although these type of rules may have exceptions, the SMR technique may take advantage of the morphological information.

POS are extracted using language analyzers, which generally count on morphological dictionaries and grammars. They do not have a unique correspondence, so one word may have different POS associated with it (e.g. *book* may be a *Noun* or a *Verb*).

These analyzers incorporate several methods of smoothing and of handling unknown words. Therefore, there is a POS for each word, and so words have always an associated word class.

7.2.3 Class Assessment

	Training		Test	
	+	-	+	-
Statistical	No supervised info	No language related info	Learn short seq	Long seq may be sparse
				Unclassified words
Morphological	Language related info	Supervised info	Learn short seq	Long seq may be sparse
			All words classified	

Table 7.1: *Class assessment.*

To summarize, Table 7.1 shows the main advantages and disadvantages of the classes presented above. Classes are evaluated by two criteria: how they are trained and how they adapt to the SMR system requirements.

7.3 Coupling SMR and SMT

The SMR approach has been designed as a translation preprocessing. Therefore, how both systems are coupled is a very important issue. The coupling strategies may differ in the training and translation steps.

7.3.1 Training Step

Here, SMR and SMT are sequentially computed. First, the SMR system is trained. Then, the training source corpus is reordered by the SMR module. Finally, the reordered source corpus is used to train the SMT system. The complete process is shown in Figure 7.7. The main advantage of using SMR in training is the reduction of the unit vocabulary sparseness.

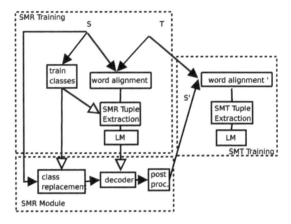

Figure 7.7: *Block diagram of the SMR and SMT training.*

New issues in SMT training

- Instead of using the source corpus, SMT uses the reordered source corpus.

- Links from the word alignment matrix tend to be ordered on a growing diagonal. Once the source corpus has been reordered, the word alignment links that were computed with the $S2T$ corpus may be kept to save computational time instead of realigning again with the $S'2T$ corpus. Translation results comparing an SMT built with the S2T word alignment or the S'2T word alignment do not vary much when using the best configuration of the SMR technique. Although the alignment quality does not change, what makes a difference is that the source text changes after applying SMR and, therefore, the alignment matrix changes. As an example, note the difference in Table 7.2, which shows the $S2T$ (left) and $S'2T$ (right) word alignment and bilingual units.

- The SMT system (except for the word alignment) is trained on the $S'2T$ (reordered-source-to-target) task. Although the links from word alignments in Table 7.2 (left and right) remain the same, the extracted units change. In general, this modification in the word alignment matrix has more impact on the unit extraction of the Ngram-based system (because of the unique segmentation) than on the phrase extraction of the phrase-based system. Recall unit extraction in § 2.3.2 and in § 2.3.1.

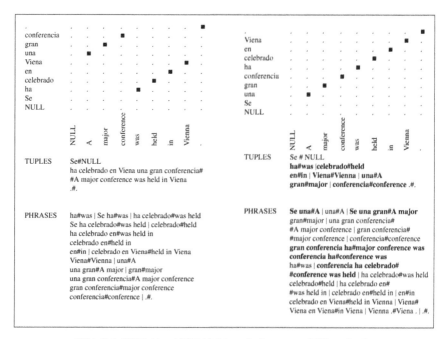

Table 7.2: *S2T (left) and S'2T (right) word alignment and bilingual units*

7.3.2 Translation Step

The block diagram of the SMR and SMT translation is shown in Figure 7.8. When translating, SMR is performed following the SMR module steps (see § 7.1.2). Then, the SMR and SMT systems are coupled using either a single-best or a graph (explained in the next lines). Finally, the SMT system performs the final translation.

7.3.2.1 Single Best (SMR-1best)

The best output of the SMR system is directly input to the SMT system. This coupling is oriented to translation tasks which do not require many reorderings.

Here, the SMT decoder may perform a monotonic search or a non-monotonic search using the maxjumps technique, see § 5.1.1. When using a monotonic search, reordering does not increase the computational cost.

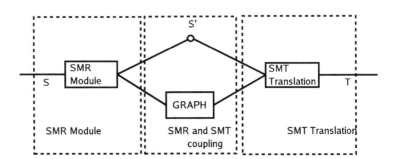

Figure 7.8: *Block diagram of the SMR and SMT translation.*

7.3.2.2 Graph (SMR-Graph) or Weighted Graph

The SMR technique generates an output graph that is introduced as an input graph for the SMT system. The weights of the reordering graph are the probabilities given by the SMR model.

Each arch of the SMR graph contains an SMR unit, which has a score from the SMR model. This SMR output graph is transformed into a word permutation graph that will be used as SMT input graph. In the case that the SMR units contain several words, this arch is extended to N arches while keeping the same path probability, (N is the number of words that contain the SMR unit). Therefore, the final input graph that is given to the SMT system is strictly a permutation of source words.

If the SMR system uses the SMR model probabilities to perform the reordering between the Source to the Reordered Source, these probabilities are kept in the graph and included as a reordering feature in the log-linear SMT framework (hereinafter, SMR-Graph$_R$). More features functions may be included (see § 7.1.1.1).

When using the graph to couple both systems, the SMT system performs a non-monotonic decoding search. Therefore, the SMT feature functions contribute to the final reordering search.

7.4 Evaluation Framework

This section reports the experimental work that evaluates the performance of the SMR approach. We have used different tasks in order to bolster our conclusions.

We carried out preliminary experiments to assess the role of word classes instead of words themselves. We compared the performance of statistical vs morphological classes. We ran core experiments to evaluate the SMR graph, features functions and efficiency performance. Additionally, we applied the SMR approach in a phrase-based system. SMR developing experiments took a monotonic baseline system as a reference. Finally, with the best SMR configuration, we performed a comparative experiment between the SMR and the lexicalized word reordering implemented in the Moses system.

7.4.1 Preliminary Experiments: Word Classes Performance vs. Lexical Words

We performed these experiments using: (1) a subset of the parallel Basic Traveler Expressions Corpus (BTEC) distributed in the 2005 IWSLT (§ A.1); and (2) the material provided by the 2005 Workshop on Machine Translation (WMT) (§ A.2). The tests sets were the ones from their respective official evaluations. Translation tasks were Chinese-to-English (Zh2En) and Spanish-to-English (Es2En), respectively.

The main goal of these preliminary experiments was to evaluate the performance of the SMR-1best approach in an Ngram-based system. Experiments are conclusive concerning the effect of including word classes in the SMR system.

7.4.1.1 System Description

The following lines report the baseline parameters of the Ngram-based system.

Preprocessing. For Chinese-to-English, Chinese preprocessing included re-segmentation using ICTCLAS (Zhang et al., 2003). For the Spanish-English pair, standard tools were used for tokenizing and filtering.

Word alignment. After preprocessing the training corpora, word-to-word alignments were performed in both alignment directions using Giza++ (Och and Ney, 2003), and the union set of both alignment directions was computed.

Tuple modeling. Tuple sets for each translation direction were extracted from the union set of alignments. The resulting tuple vocabularies were pruned by considering the N best translations for each tuple source-side ($N = 20$ for the Spanish-to-English task and no limit for the Chinese-to-English task) in terms of occurrences. The SRI language modeling toolkit (Stolcke, 2002)[2] was used to compute the bilingual 4-gram language model. Kneser-Ney smoothing (Kneser and Ney, 1995) and interpolation of higher and lower n-grams were used to estimate the 4-gram translation language models.

Features functions. Lexicon models were used in the source-to-target and target-to-source directions. A word bonus and a target language model were added in decoding. Again, Kneser-Ney smoothing (Kneser and Ney, 1995) and interpolation of higher and lower n-grams were used to estimate the 4-gram target language models.

Optimization. Once models were computed, optimal log-linear coefficients were estimated for each translation direction and system configuration using an in-house implementation of the widely used downhill Simplex method (Nelder and Mead, 1965) (using the single-loop detailed in §2.5). The BLEU score was used as the objective function.

Decoding. The decoder was set to perform histogram pruning, keeping the best $b = 50$ hypotheses (during the optimization work, histogram pruning is set to keep the best $b = 10$ hypotheses). Spanish-to-English task considered only a monotone search, while Chinese-to-English considered a monotone and a non-monotone search. When allowing for reordering in the SMT decoder, the distortion (m) and reordering limit (j) (see § 2.3.2) were empirically set to $m = 5$ and $j = 3$, as they showed a good trade-off between quality and efficiency.

The following reports the SMR default parameters.

SMR parameters. The bilingual unit extraction did not have any limit for unit lengths. The resulting tuple vocabularies were pruned by considering the N best translations for each tuple source-side ($N = 30$ for all translation tasks). The SMR Model were a 5gram and 4gram (in the WMT and BTEC task, respectively) back-off language model with Kneser-Ney smoothing and were built with the SRILM. Decoding involved a beam search of 50. All experiments in this section were carried out using either 100 statistical classes or no word classes to perform the SMR-1best.

7.4.1.2   SMR-1best and Statistical Classes

Statistics. To begin, the statistics of both approaches ($S2T$ and $S'2T$) are shown for the Zh2En task. Table 7.3 shows the vocabulary of bilingual n-grams and embedded words in the translation model. Once the reordering translation is computed, the word alignment matrix is more monotonic. As stated in 2.3.2 non-monotonicity poses difficulties in tuple extraction. Therefore,

---

[2]http://www.speech.sri.com/projects/srilm/

when word alignment becomes more monotonic (as shown in Figure 7.2), we expect a reduction in the sparseness of the tuple vocabulary, and, therefore an improvement in the quality of translation. Table 7.3 confirms a significant enlargement of the number of translation units, which leads to a growth of the translation vocabulary. We also observe a decrease in the number of embedded words (around 20%). From § 2.3.2, we know that the probability of embedded words is estimated independently of the translation model. Reducing embedded words allows for a better estimation of the translation model.

System	1gr	2gr	3gr	4gr	Embedded
NB	34487	57597	3536	1918	5735
SMR + NB	35638	70947	5894	3412	4632

Table 7.3: *Vocabulary of n-grams and embedded words in the translation model.*

Table 7.4 shows the tuples used to translate the test set (total number and vocabulary). Note that the number of tuples and vocabulary used to translate the test set is significantly greater after SMR.

Preliminary translation results. The proposed SMR approach was firstly evaluated using a monotonic search (NB or monotonic baseline configuration) and a non-monotonic search (NBnm or non-monotonic baseline configuration).

Table 7.5 shows the results for the Zh2En task (NB and NBnm) and for the Es2En task (NB).

Using SMR (without word classes) in a monotonic search increases BLEU in Zh2En and Es2En by 0.3 and 1 points. respectively. Using statistical classes instead of words themselves gives an improvement of 1 and 2.2 points BLEU for Zh2En and Es2En, respectively. Finally, using a non-monotonic search yield an improvement of 2 points BLEU for Zh2En. We did not used it in Es2En because the local constrains imply a too high computational cost and do not generalize well (Crego and Mariño, 2007).

System	Total	Vocabulary
NB	4460	959
SMR + NB	4628	1052

Table 7.4: *Tuples used to translate the test set (total number and vocabulary).*

System	Classes	BLEU	NIST	mWER	mPER
Zh2En					
NB	-	42.42	8.3	42.87	33.44
NBnm	-	43.58	8.9	43.89	34.05
SMR + NB	-	42.73	8.9	45.26	34.82
SMR + NB	100	43.75	8.49	42.45	33.85
SMR + NBnm	100	**45.97**	**9.0**	**40.92**	**32.32**
Es2En					
NB	-	27.69	7.31	61.6	45.34
SMR + NB	-	28.60	7.53	59.89	43.53
SMR + NB	100	**30.89**	**7.75**	**55.77**	**42.85**

Table 7.5: *Results in the test set of the Zh2En and Es2En tasks.*

7.4.1.3 Preliminary Conclusions

Both BLEU and NIST coherently increase with the inclusion of the SMR step when statistical classes are used. The improvement in translation quality can be explained as follows:

- The new task $S'2T$ becomes more monotonic. Therefore, there is a reduction in the sparseness of the translation units which tend to be shorter.

- SMR takes advantage of the use of classes by allowing to infer new reorderings. The use of classes helps to successfully capture word reorderings that are missed in the standard SMT monotonic system.

The gain obtained in the SMR+NBnm case indicates that the reordering provided by the SMR system and the non-monotonic search are complementary. The SMR output could still be further monotonized.

7.4.2 Statistical and Morphological Word Classes Comparison

This section evaluates several types of word classes and n-gram lengths in the SMR model in order to choose the SMR configuration that provides the best translation in terms of quality. We have designed an experiment in order to accomplish this evaluation without building an SMT system. Given 500 manually aligned parallel sentences of the Epps corpus (Lambert et al., 2006), we order the source test in the way that better matches the target set. This ordered source

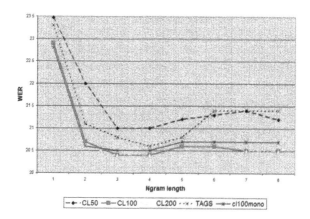

Figure 7.9: *WER over the reference given various sets of classes and n-gram lengths.*

set is considered our reference as it is based on manual alignments. Additionally, the 500 sentences set is also translated using the SMR configurations to be tested. Finally, the mWER is used as quality measure.

The manually aligned corpora is only available in the Spanish-English pair. Therefore, experiments were only carried out using the parallel corpus released in the 2007 WMT (§ A.2) in Spanish to English.

Figure 7.9 shows the WER behavior given different types of classes. For statistical classes (*cl50,cl100,cl200*) we used the Och monolingual classes (Och, 1999), that can be obtained using *mkcls* [3]. We also used the statistically reordered classes (*cl100mono*) which were explained in Chapter 7.2. Both statistical and statistically reordered classes used the *disamb* tool of SRILM to classify unknown words. As morphological classes we used the POS provided by the FreeLing[4] tool (Carreras et al., 2004). Spanish words were classified into the following categories: adjectives, adverbs, determinants, nouns, verbs, pronouns, conjunctions, interjections, prepositions, punctuation and numerals. Within these categories, there were other attributes related to the categories, i.e., qualifying or ordinal for an adjective. Additionally, number or gender information is included when required. Taking into account this morphological information, there were 332 different tags in the Spanish corpus.

The results show that statistical classes perform better than POS tags and the best results

---

[3]http://www.fjoch.com/mkcls.html
[4]http://www.lsi.upc.edu/~nlp/freeling/

SRC: *el establecimiento de **vínculos más estrechos** entre la **Unión Europea** y*
Croacia dependen de **manera decisiva**
SMR OUTPUT TAGS: *el establecimiento de vínculos más estrechos entre la Europea Unión y*
Croacia dependen de manera decisiva
SMR OUTPUT CL50: *el establecimiento de más estrechos vínculos entre la Europea Unión y*
Croacia dependen de manera decisiva
SMR OUTPUT CL100: *el establecimiento de más estrechos vínculos entre la Europea Unión y*
Croacia dependen de decisiva manera
SMR OUTPUT CL200: *el establecimiento de más estrechos vínculos entre la Europea Unión y*
Croacia dependen de decisiva manera
SMR OUTPUT CL100ᴍᴏɴᴏ: *el establecimiento de más estrechos vínculos entre la Europea Unión y*
Croacia dependen de manera decisiva
REFERENCE: *el establecimiento de **más estrechos vínculos** entre la **Europea Unión** y*
Croacia dependen de **decisiva manera**

Figure 7.10: *Example of SMR output sentences using: morphological (TAGS) or statistical (cl50, cl100, cl200, cl100mono) classes.*

can be achieved with 100 and 200 classes and an n-gram length of 5. Using a 4gram language model, the biggest difference (0.6 WER) is between 50 and 100 (or 200) statistical classes. Apart from that, the difference between using POS tags or 100 statistical classes is only 0.2 WER, which may not generalize to other tasks.

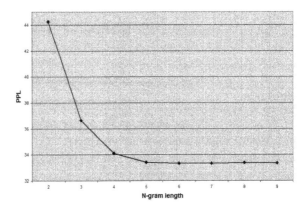

Figure 7.11: *Perplexity over the manually aligned test set given the SMR Ngram length.*

7.4.3   Core Experiments: Coupling, Features, Reorderings and Efficiency.

This section reports the main experiments performed with the SMR approach in an Ngram-based SMT system. After defining the system, the SMR-Graph, the SMR feature functions and the SMT efficiency parameters were evaluated.

Experiments were carried out using the parallel corpus developed (1) in the TC-STAR project (§ A.3) (Spanish to English), (2) in the 2007 WMT evaluation (§ A.2) (German to English) and (3) in the 2007 IWSLT evaluation (§ A.1) (Arabic to English). The tests sets were the ones used for the 2nd Tc-Star evaluation, for the 2007 WMT News evaluation and for the 2007 IWSLT evaluation, respectively.

7.4.3.1 System Description

The word alignment and language model computation are the same as reported in §7.4.3.1.

Preprocessing. For the Spanish-to-English task, standard tools were used for tokenizing and filtering. The English side of the training corpus was POS tagged using the freely available TnT[5] tagger (Brants, 2000), while for the Spanish side we used the freely available FreeLing[6] tool (Carreras et al., 2004).

For the German-to-English task, standard tools were used for tokenizing and filtering. Again, the English side of the training corpus was POS tagged using the TnT tagger.

For the Arabic-to-English task, Arabic tokenization was performed following the Arabic TreeBank tokenization scheme: 4-way normalized segments into conjunction, particle, word and pronominal clitic. Tokenization was performed using the publicly available Morphological Analysis and Disambiguation (MADA) tool (Habash and Rambow, 2005) together with TOKAN, a general tokenizer for Arabic. English preprocessing simply included down-casing, separating punctuation from words and splitting off 's. The English side is POS-tagged with the TnT tagger.

Tuple modeling. Tuple sets for each translation direction were extracted from the union set of alignments. The resulting tuple vocabularies were pruned by considering the N best translations for each tuple source-side ($N = 20$ for the Spanish-to-English direction, $N = 30$ for the German-to-English direction and $N = 40$ for the Arabic-to-English direction) in terms of occurrences. A 4-gram translation model was used.

Features functions. Lexicon models were used for the source-to-target and target-to-source directions. A word bonus, a 4-gram target language model and a 5-gram POS target language model were added in decoding.

Optimization. Once models were computed, optimal log-linear coefficients were estimated

---

[5] http://www.coli.uni-saarland.de/~thorsten/tnt/
[6] http://www.lsi.upc.edu/~nlp/freeling/

for each translation direction and system configuration using an in-house implementation of the widely used downhill Simplex method (Nelder and Mead, 1965) (using the double-loop detailed in §2.5). The BLEU score was used as objective function.

Decoding. The decoder was set to perform histogram pruning, keeping the best $b = 50$ hypotheses (during the optimization work, histogram pruning is set to keep the best $b = 10$ hypotheses). Notice that the NB baseline system is monotonic.

SMR default parameters. The bilingual unit extraction did not have any limit for unit lengths. The resulting tuple vocabularies are pruned by considering the N best translations for each tuple source-side ($N = 30$ for all translation tasks) The SMR Model was a 5gram back-off language model with Kneser-Ney smoothing and was built with the SRILM. As default, 100 statistical word classes were used for the Spanish to English pair; 200 statistical word classes for the German to English tasks; and 100 statistical classes were used for the Arabic-to-English task. The beam search was set to 5.

7.4.3.2 Experiments in Coupling SMR and SMT

The SMR system may be coupled with the SMT system either using a single best or a graph, see § 7.3. As it could be expected, experimental results show that the latter performs much better given that it offers several reordering hypotheses.

Table 7.6 presents the BLEU, NIST, mPER and mWER scores to compare the NB and SMR+NB approaches. The SMR approach improves all measures, especially when using the graph approach.

Additionally, Table 7.6 provides the number of words and translation time. The SMR-1best option keeps translation time similar to the baseline system. Whereas, there is a moderate increase of the computational cost when using the SMR-Graph.

Figure 7.12 shows typical examples of translated sentences, where the NB baseline system is compared to the SMR+NB system. Es2En language pair usually requires local reorderings, like noun plus adjective that swaps from one language to the other. German has traditionally been considered problematic because of the position of the verb, which is in second position in a main clause and at the end in a subordinate clause. De2En examples report reorderings involving up to five words that handle this change in verb position. Finally, in general the Ar2En task again tends to present local reorderings. Additionally, given the touristic domain of the Ar2En BTEC task, test sentences are not very long. Examples show local reorderings either in question or affirmative sentences.

System	BLEU	NIST	PER	WER	Words	Time
Es2En						
NB	52.57	10.64	26.63	36.97	29.2k	50.0'
SMR-1best + NB	52.95	10.62	26.84	36.96	29.4k	52.8'
SMR-Graph + NB	53.72	10.72	26.43	36.37	29.4k	112.7'
SMR-Graph$_R$ + NB	54.51	**10.81**	26.24	**35.67**	29.0k	78.7'
SMR-Graph$_T$ + NB	54.11	**10.81**	**26.11**	35.82	29.3k	129.2'
SMR-Graph$_{R+T}$ + NB	**54.54**	**10.81**	26.26	**35.65**	29.3k	122.8'
De2En						
NB	21.30	6.69	47.90	65.61	50.5k	85.9'
SMR-1best + NB	21.61	6.82	46.80	64.79	48.6k	76.9'
SMR-Graph + NB	21.89	6.89	46.10	64.39	49.9k	77.0'
SMR-Graph$_R$ + NB	23.08	6.96	46.25	63.60	49.9k	125'
SMR-Graph$_T$ + NB	22.18	6.93	46.33	63.80	48.5k	110.6'
SMR-Graph$_{R+T}$ + NB	**23.30**	**7.02**	**46.10**	**62.98**	49.0k	120'
Ar2En						
NB	45.00	7.65	34.92	39.15	3.6k	3.1'
SMR-1best+NB	46.45	7.90	32.32	36.73	3.6k	2.1'
SMR-Graph+NB	48.58	8.02	31.15	34.76	3.5k	5.8'
SMR-Graph$_R$+NB	48.01	8.01	31.38	35.21	3.6k	5.7'
SMR-Graph$_T$ + NB	48.47	8.02	31.41	35.08	3.6k	5.5'
SMR-Graph$_{R+T}$ + NB	**49.35**	**8.13**	**30.70**	**34.23**	3.5k	6.5'

Table 7.6: *Translation results and computational time for several couplings of the SMR and SMT (including none, one or two features functions).*

NB: not only through a compromise economic immediately
SMR+NB: not only through an **immediate economic commitment**
REF: not only through an immediate economic commitment
NB: The European Union must be a political element essential for the fight against terrorism
SMR+NB: The UE should be an **essential political element** in the fight against terrorism
REF: The UE must be an essential political element to fight against terrorism
NB: The Group of the European Peoples has asked (...)
SMR+NB: The **European Peoples Group has requested (...)**
REF:The European Popular Group asked (...)
NB: Iraq needs several years a new constitution to write
SMR+NB: Iraq needs several years **to write a new constitution,**
REF: Iraq needs several years to write a new constitution,
NB: EU membership has result in a state decisive measures must accept
SMR+NB: the EU membership result, a state **must accept radical measures,**
REF: EU membership entails having to accept incisive measures.
NB: (..) that the death penalty threatened murderers go further would her arrest to escape,
SMR+NB: (..) that the death penalty threatened murderers **would go even further,**
REF: (..) that capital punishment may make a murderer fight harder
NB: Broke one of them room.
SMR+NB: Someone **broke them our room.**
REF: Someone broke into our room.
NB: Can you discount a little it?
SMR+NB: Can you discount **it a little?**
REF: Can't you lower the price?
NB: I'm sorry but not this what I think.
SMR+NB: I'm sorry **but this is not what I think.**
REF: I'm sorry but this is not what I have in mind.

Figure 7.12: *Translation examples from the* NB *and* SMR+NB *systems: Es2En, De2En and Ar2En (from top to bottom).*

7.4.3.3 SMR with Additional Feature Functions

The reordering graph may be used without weights (SMR-Graph) or with weights. The weights of the reordering graph are the probabilities given by the SMR Model (SMR-Graph$_R$) and/or the SMR target language model (SMR-Graph$_{R+T}$/SMR-Graph$_T$). The former may be seen as the equivalent of the translation model in a SMT system. The latter may be seen as a reordered source language model. Both were introduced in § 7.1.2. Experiments were performed for the same tasks as above.

Table 7.6 shows the SMR features performance. The more reordering models used to weight the reordering hypothesis, the better results obtained. One main reason for that is the weights of the reordering models are optimized together with the translation feature functions to improve the translation score.

Reordering	Frequency
1 0	1020
1 2 0	171
2 1 0	36
1 2 3 0	25
2 3 0 1	16
1 3 2 0	11
3 0 1 2	8
2 3 4 0 1	4
1 2 3 4 5 0	3

Table 7.7: *Most frequent reorderings performed in the Es2En test set.*

Under these circumstances, Table 7.7 shows the most frequent reorderings that have been performed in the test set. Considering the Es2En test set, there are more than 2,000 reorderings and a vocabulary of 36 reorderings. There is no reordering limit and a reordering of up to 9 words has been performed. See the example in Figure 7.13.

S: LOGRAR EN LA REGIÓN UNA PAZ JUSTA Y DURADERA
S': JUSTA Y DURADERA UNA PAZ LOGRAR EN LA REGIÓN
T: FAIR AND LASTING PEACE IN THE REGION
Ref: A FAIR AND LASTING PEACE IN THE REGION

Figure 7.13: *Example of long reordering: source (S), reordered source (S'), translation (T) and reference (R)*

7.4.3.4   SMR Performance in a Ngram-based System

An analysis of the translation units is reported both in the baseline SMT system (NB) and the baseline SMT system enhanced with the SMR system (SMR+NB).

As stated in the preliminary experiments the source training reordering allows for fewer crossings in word alignments. The non-monotonicity poses difficulties for units extraction in an Ngram-based system. Figure 7.14 shows an example of how SMT units are modified when using the SMR approach as preprocessing in SMT training. Clearly, the SMT units are reduced in length. As a consequence, there is a reduction in SMT vocabulary shown in Table 7.8. More-

NB:	hablar en esta Asamblea de manera provechosa e interesante#useful and interesting discussions in this House
SMR+NB:	de#NULL | manera#NULL | provechosa#useful | e#and | interesante#interesting | hablar#discussions | en#in | esta#this | Asamblea#House
NB:	aus der Presse und dem Fernsehen wissen#aware from the press and television
SMR+NB:	wissen#aware | aus#from | der#the | Presse#press | und#and | dem#NULL | Fernsehen#television
NB:	hl#is | ywjd#there | any grfp b+ sEr >rxS#cheaper room
SMR+NB:	hl#is | ywjd#there any | b+#NULL sEr >rxS#cheaper | grfp#room

Figure 7.14: *Examples of tuples extracted from a training sentence pair in the baseline (NB) and in the enhanced (SMR+NB) system. '|' marks unit boundaries and '#' marks the two components of a bilingual unit. Arabic is written in Buckwalter.*

over, with this enhanced approach, we observe a decrease in the number of embedded words. Reducing embedded words allows for a decrease in the OOV words in translation.

7.4.3.5 Additional Experiments: SMR Efficiency Parameters

The more complex is the reordering graph given to the SMT system, the less efficient the translation. This section reports several beam searches used in the SMR decoding. Additionally, the impact of the SMR main feature is reported for each pruning.

Experiments were carried out with a similar baseline SMT system as described in § 7.4.3.1 for Spanish to English. Results are reported for both translation directions.

Given that the reordering graph is the output of a beam search decoder, the reordering graph was pruned by limiting the SMR beam, i.e., limiting the size of hypothesis stacks.

Given a reordering graph, another option is to prune states and arches only used in paths s times worse than the best path.

Table 7.9 gives the results of the proposed pruning. Note that computational time is given in terms of the monotonic translation time. It is clear that graph pruning guarantees efficiency for the system and even increases the translation's quality. Similar results are obtained in terms of BLEU for both types of pruning. In this task, it seems more appropriate to limit the beam search directly when performing the SMR reordering instead of pruning the graph separately.

As expected, the SMR Model weight has a bigger impact when limiting the graph pruning (i.e., when the graph has more paths).

	NB	SMR+NB	Relative increment
Es2En			
1-word tuples	439.6k	507.5k	15.4%
>1-word tuples	1.8M	1.3M	-36%
Embedded words	57.1k	46.6k	-13%
De2En			
1-word tuples	618.1k	711k	15%
>1-word tuples	2.1M	1.7M	-25%
Embedded words	211.8k	163.5k	-30%
Ar2En			
1-word tuples	9.5k	10.3k	8%
>1-word tuples	20.4k	18.8k	-8%
Embedded Words	40.0k	34.5k	-16%

Table 7.8: *Variation in the size of the translation vocabulary (1-word and longer than 1-word tuples).*

System	Pruning	$BLEU_{En2Es}$	$BLEU_{Es2En}$	TIME
NB	-	30.50	30.20	T_m
SMR-Graph$_R$ + NB	**b5**	**31.32**	**32.64**	$2.4T_m$
SMR-Graph + NB	b5	31.25	31.82	$2.5T_m$
SMR-Graph$_R$ + NB	b50	30.95	32.28	$5.3T_m$
SMR-Graph + NB	b50	30.90	27.44	$4.8T_m$
SMR-Graph$_R$ + NB	b50 s10	31.19	32.20	$1.5T_m$
SMR-Graph + NB	b50 s10	31.07	32.41	$\mathbf{1.4T_m}$

Table 7.9: *BLEU performance experimenting: different graph prunings (b stands for beam and s for states); the influence of the SMR Model weight. The computational time is reported related to T_m (monotonic translation time).*

7.4.4   SMR in a Phrase-based System

This section describe experiments on the SMR behavior in a state-of-the-art phrase-based SMT system.

Experiments were carried out using the parallel corpus developed (1) in the TC-STAR project (§ A.3) using the 2nd evaluation official test set, and (2) in the IWSLT evaluation (§ A.1), using the 2007 IWSLT official test set. Tasks were Spanish-to-English and Arabic-to-English, respectively.

7.4.4.1   System Description

The preprocesssing, word alignment, optimization, decoding and SMR parameters used in the phrase-based systems are the same as those reported in §7.4.3.1.

Phrase modeling. Phrase sets for each translation direction were extracted from the union set of alignments. The resulting phrase vocabularies were pruned considering the N best translations for each tuple source-side ($N = 20$ for the Spanish-to-English direction and $N = 40$ for the Arabic-to-English direction) in terms of occurrences. Phrases up to a length of 10 were considered.

Features functions. The conditional and posterior probabilities were used as translation models. Lexicon models were used in the source-to-target and target-to-source directions. A word and phrase bonus, a 4-gram target language model and a 5-gram POS target language model were added in decoding.

Decoding. Again, the decoder was set to perform histogram pruning, keeping the best $b = 50$ hypotheses (during the optimization work, histogram pruning is set to keep the best $b = 10$ hypotheses). Notice that the PB baseline system is monotonic.

7.4.4.2   SMR in a Phrase-based System

Table 7.10 shows that the SMR can successfully be applied to a phrase-based system. Given the best SMR configuration from §7.4.3.2, the SMR+PB system provides an improvement in BLEU of 2.2 on Es2En and 3.3 points on Ar2En.

Although the effect of monotonizing the source training may have less influence in a phrase-based system than in an Ngram-based (because of the difference in unit extraction), the SMR approach has almost the same impact on both systems.

System	BLEU	NIST	PER	WER	Words	Time
Es2En						
PB	51.48	10.54	26.95	37.85	30.8k	101.5'
SMR-Graph$_{R+T}$+PB	**53.70**	**10.75**	**26.37**	**36.00**	29.1k	310.6'
Ar2En						
PB	43.70	7.62	32.87	37.40	3.5k	1.6'
SMR-Graph$_{R+T}$+PB	**47.00**	**7.81**	**31.44**	**34.88**	3.4k	4.1'

Table 7.10: *Translation results, number of translation words and computational time for the PB system and for the SMR+PB system.*

System	BLEU	NIST	PER	WER	Words	Time
Es2En						
Moses + LR	53.59	10.80	**25.74**	35.75	28.8k	264,3'
SMR-Graph$_{R+T}$ + PB	53.70	10.75	26.37	36.00	29.1k	310.6'
SMR-Graph$_{R+T}$ + NB	**54.54**	**10.81**	26.26	**35.65**	29.3k	122.8'
Ar2En						
Moses + LR	48.57	7.78	**30.83**	**34.21**	3.3k	3.8'
SMR-Graph$_{R+T}$ + PB	47.00	7.81	31.44	34.88	3.4k	4.1'
SMR-Graph$_{R+T}$ + NB	**49.35**	**8.13**	30.70	34.23	3.5k	6.5'

Table 7.11: *SMR in a phrase- and Ngram-based system compared to the lexicalized reordering implemented in Moses.*

7.4.5 Comparison of SMR to other State-of-the-art Reordering Approaches

Here, two strategies to compare the SMR approach against other state-of-the-art reordering techniques are reported: the results in several international evaluation campaigns § B and the performance of the SMR approach against the lexicalized reordering implemented in Moses [7].

Again, experiments were carried out using the same tasks and test sets as in §7.4.4.

7.4.5.1 Reordering Comparison

Table 7.6 provides a comparison between our approach and the lexicalized reordering (LR) method (Tillmann, 2004; Koehn et al., 2005) used in the open source Moses toolkit.

---

[7]Experiments were done with the 2006 version of Moses

Moses. The Moses system was built following the provided guidelines[8].The parameters of the lexicalized reordering were settled as follows.

- The lexicalized distortion model was defined as *msd-bidirectional-fe*. *Msd* means the reordering types can be monotone, swap and discontinuous. *Bidirectional* means that certain phrases may not only flag, if they themselves are moved out of order, but also if subsequent phrases are reordered. *fe* concerns out of sparse data and the probability distribution conditions on the foreign phrase f and on the English phrase e.

- The maximum number of words to be reordered (max-skip) was set to 6.

The preprocessing, the alignment, the feature functions (except for the reordering ones) and the decoding parameters were the same as those reported in §7.4.3.1. The reordering, decoder and translation models were the only differences between Moses and our system.

Automatic measures in Table 7.6 show that SMR+NB outperforms Moses+LR in the Es2En and the Ar2En tasks. After analyzing the translations manually, most differences were due to reorderings, which were better solved when using the SMR approach. This conclusion is supported by the values of mPER and mWER. For all tasks, the improvement in mWER is higher than the improvement in mPER. Figure 7.15 shows typical examples of translated sentences.

---

[8]http://www.statmt.org/wmt07/baseline.html

SMR+NB: not only through an **immediate economic commitment**

SMR+PB: not only through an **immediate economic commitment**

Moses: not only through a compromise immediate economic

REF: not only through an immediate economic commitment

SMR+NB: The UE should be an **essential political element** in the fight against terrorism

SMR+PB: The UE must be an **essential political element** in the fight against terrorism

Moses: The UE must be a political element essential for the fight against terrorism

REF: The UE must be an essential political element to fight against terrorism

---

SMR+NB: The **European Peoples Group has requested that replaces the debate**

SMR+PB: The **European Peoples Group has requested to replaces the debate**

Moses: The Group of the European Peoples replace has requested that the debate

REF: The European Popular Group asked to replace the debate

SMR+NB: But this summer has also been marked by other **luctuosos events**:

SMR+PB: But this summer has also been marked by other **luctuosos events**:

Moses: But this summer has also been marked by other events luctuosos:

REF: But this summer was also marked by other tragic events:

SMR+NB: the **Spanish and French States** where 50 kilometers there are **five toll barriers**

SMR+PB: the **Spanish and French States**, where in 50 kilometers there are **five toll barriers**

Moses: the States Spanish and French, where in 50 kilometers there are five barriers of toll

REF: the Spanish and the French States, where along 50 kilometers there are five toll barriers

===

SMR+NB: Someone **broke them our room.**

SMR+PB: Someone **broke them our room.**

Moses: Our room one of them.

REF: Someone broke into our room.

SMR+NB: **I'm sorry to tell you**

SMR+PB: **Accept cash.** Only cash.

Moses: Cash **only accept.**

REF: We only accept cash.

SMR+NB: Where is the **taxi waiting stand**?

SMR+PB: Where is the **taxi waiting stand**?

Moses: Where's the **taxi stand** waiting?

REF: Where is a taxi stand?

SMR+NB: **Can you discount it a little?**

SMR+PB: A **little discount.**

Moses: Discount it is in a little.

REF: A bit cheaper.

SMR+NB: **I'm sorry but this is not what I think.**

SMR+PB: **Isn't this souvenir from my friend?**

Moses: **No, this is a souvenir from my friend.**

REF: No, this is a souvenir from my friend.

Figure 7.15: *Translation examples from the SMR+NB, SMR+PB and Moses system: from the* Epps *(up) and IWSLT (down) tasks.*

7.5   Conclusions and Further Work

We have addressed the reordering challenge in SMT by using the same powerful translation techniques to generate source reordering hypotheses.

SMR allows for a reduction of the vocabulary sparseness of the Ngram-based SMT system during the training phase. Note especially that SMR reduces the embedded words whose probability can not be estimated together with the translation model that must be calculated independently. Therefore, reducing embedded words allows for a better estimation of the translation model.

Using classes to train the reordering hypothesis (instead of words themselves) allows us to generalize in the test phase. Therefore, the SMR technique is able to generate reordering hypotheses of sequences of words that were not seen during training. Additionally, the SMR technique provides a smoothed context-based weight to each reordering hypothesis by taking advantage of the highly developed language model techniques.

Although introducing reordering abilities increases the system computational cost, experiments show that using the SMR technique guides the final translation decoding in an efficient manner.

Results have been presented in 4 languages pairs (ZhEn, EsEn, DeEn and ArEn). Reordering with the SMR technique highly outperforms our baseline system. SMR has also been compared against the lexicalized reordering method used in the open source Moses toolkit. The results show improved performance on the Es2En and Ar2En tasks.

The main drawback of this approach is that although word classes are able to generalize reorderings, long distance reorderings are not usually performed in the test set. Manual analysis shows that long SMR hypotheses may tend to be sparse. In order to solve this main drawback, the SMR system may add syntactical information to learn longer reorderings better. An even better option would be to develop the SMR system using a hierarchical system that captures longer reorderings.

RELATED PUBLICATIONS

- *An Ngram-based reordering model*
  Costa-jussà, M.R. and Fonollosa, J.A.R.
  Computer Speech and Language vol 23, num 3, issn 0885-2308, pp 362-375, 2009

- Costa-jussà, M.R. and Fonollosa, J.A.R.
  Analysis of Statistical and Morphological Classes to Generate Weigthed Reordering Hypotheses on a Statistical Machine Translation System
  Second Workshop of Statistical Machine Translation (WMT07) ACL, pp 171-176, Prague, July 2007

- Costa-jussà, M.R., Crego, J.M., de Gispert, A., Lambert, P., Khalilov, M., Fonollosa, J.A.R., Mariño,J-B,and Banchs, R.
  Ngram-Based Statistical Machine Translation Enhanced with Multiple Weighted Reordering Hypotheses Second Workshop of Statistical Machine Translation (WMT07) ACL, pp 167-170, Prague, July 2007

- Costa-jussà, M.R., Crego, J.M., de Gispert, A., Lambert, P., Khalilov, M., Fonollosa, J.A.R., Mariño,J-B,and Banchs, R.
  TALP Phrase-based Statistical Machine Translation and TALP system combination the IWSLT 2006 International Workshop on Spoken Language Translation (IWSLT), pp 123-129, Kyoto November 2006.

- Costa-jussà, M.R. and Fonollosa, J.A.R.
  Sistema de Reordenamiento Estadístico
  Procesamiento del Lenguaje Natural (SEPLN06), pp 249-256, Zaragoza, Septiembre 2006.

- Costa-jussà, M.R. and Fonollosa, J.A.R.
  Statistical Machine Reordering
  Empirical Methods in Natural Language Processing (EMNLP06), pp 70-77, Sydney July 2006.

8

Conclusions

This PhD extends the state-of-the-art in SMT especially by developing new statistical re-ordering models. Moreover, new techniques have innovatively and successfully been applied to statistical machine translation systems. The main scientific contributions following a chronological order can be summarized as follows:

Introduction of continuous language models in an Ngram-based system. The basic idea of continuous space language models is to project the word indices onto a continuous space and to use a probability estimator operating on this space. Since the resulting probability functions are smooth functions of the word representation, better generalization to unknown n-grams can be expected. The first contribution of this thesis is the introduction of continuous space language models to smooth the bilingual translation language model of an Ngram-based system. Generalization to unseen events may not be very important in an Ngram-based translation models since the system searches only for the best matching translation unit pair among the existing ones. The main advantage is that in a continuous space language model the posterior probabilities are *interpolated* for any possible context of length $n - 1$ instead of backing-off to shorter contexts. Additionally, the resulting probability function of the continuous space translation model is better estimated than in other standard smoothing methods given the significant reduction in perplexity. Therefore, translation units are more adequately chosen and the translation quality is improved. Furthermore, continuous space language models were applied to recalculate the probabilities of the target language model. Again the translation quality is improved as shown by the higher BLEU scores and judging by human criteria the introduction of continuous space language models to compute the target language model seems to be improving the translation fluency. When both techniques (continuous space translation and target language model) are applied together individual improvements almost add up.

Development of a phrase- and Ngram-based system combination. The phrase- and Ngram-based SMT systems have been further developed and analyzed through experimentation with different tasks. This development and analysis has lead to the implementation of the two-system combination using the score given by both SMT systems. The concatenation of two-system outputs increases the translation alternatives and the introduction of the probability given by the corresponding and the opposite system allows the appropriate choice among the hypotheses.

Introduction of novel statistical reordering techniques. Two novel statistical reordering algorithms have been proposed aimed at solving the reordering challenge in an SMT system for any language pair. Both are based on the extended theory that *the translator will undo the syntactic structure of the source text and then formulate the corresponding message in the target language* [1]. Our approaches try to undo the syntactic structure of the source text in order to better match the target text. This is expanded below:

- The **Recursive Alignment Block Classification** technique is able to cope with swapping examples seen during training; it manages to infer properties that might allow reordering in pairs of blocks (sequences of words) not seen together during training; and, finally, it can present robustness with respect to training errors and ambiguities. These objectives are achieved by learning statistical reordering rules from the training set and inferring new reordering rules without requiring additional knowledge. This reordering technique provides more accurate local reorderings than those provided only by translation units.

- The **Statistical Machine Reordering** (SMR) approach consists of converting the entire source corpora into an intermediate representation, in which source-language words are presented in an order that more closely matches that of the target language. To infer new reorderings that were not learned during training, the developed approach makes use of

---

[1] http://accurapid.com/journal/05theory.htm

word classes. To correctly integrate the SMR and SMT systems, both are concatenated by using a word graph which offers weighted reordering hypotheses to the SMT system.

SMR allows for a reduction of the vocabulary sparseness of the Ngram-based SMT system during the training phase and, as a consequence, the translation units are better estimated.

Using classes to train the reordering hypothesis (instead of words themselves) allows the inference of new reordering hypotheses.

In addition, each reordering hypothesis is scored by taking advantage of the highly developed language model techniques. This information increases the probability of choosing an adequate reordering.

Although introducing reordering abilities increases the system computational cost, experiments show that using the SMR technique guides the final translation decoding in an efficient manner.

Reordering with the SMR technique significantly outperforms our baseline system. SMR has also been compared against the lexicalized reordering method used in the open source Moses toolkit, and the SMR technique shows very competitive results.

Corpora Description

This appendix puts together all the corpus sets which have been used to perform this thesis' experiments. Corpora were generally taken from a project framework or an evaluation campaign. Therefore, the internal experiment performance can be easily contrasted.

The training data is preprocessed by using standard tools for tokenizing and filtering. In the filtering stage, sentence pairs are removed from the training data in order to allow for a better performance of the word alignment tool. Sentence pairs are removed according to two criteria:

- Fertility filtering: removes sentence pairs with a word ratio larger than a predefined threshold value.

- Length filtering: removes sentence pairs with at least one sentence of more than 100 words in length. This helps to maintain alignment computational times bounded.

Next we detail the corpora used all along this thesis work. Tables present the basic statistics for the training, development and test data sets for each considered language. More specifically, the statistics show the number of sentences, the number of words, the vocabulary size (or number of distinct words) and the number of available translation references.

A.1   IWSLT data

Data employed here consists in sentences randomly selected from the BTEC (Basic Travel Expression Corpus) corpus (Takezawa et al., 2002). Development and test sets correspond to the official 2005, 2006 and 2007 IWSLT evaluation data sets[1].

Table A.1 presents the basic statistics of training, development and test data sets for the 2005 and 2006 Chinese-to-English task.

		2005 Evaluation		2006 Evaluation	
		Chinese	English	Chinese	English
Train	Sentences	20k		40.0k	
	Words	176.2k	182.3k	342.3k	367.3k
	Vocabulary	8.7k	7.0k	11.2k	7.2k
Dev	Sentences	506		489	
	Words	3.5k	-	5.7k	-
	Vocabulary	870	-	1,1k	-
	References	16			
Test	Sentences	506		500	
	Words	4k	-	6,1	-
	Vocabulary	916	-	1,3k	-
	References	16			

Table A.1: *BTEC Chinese-English corpus. Basic statistics for the training, development and test data sets.*

Table A.2 presents the basic statistics of training, development and test data sets for the 2006 and 2007 Arabic-to-English task.

Table A.3 presents the basic statistics of training, development and test data sets for the 2006 Japanese-to-English and Italian-to-English task.

---

[1]http://iwslt07.itc.it/

		2006 Evaluation		2007 Evaluation	
		Arabic	English	Arabic	English
Train	Sentences	23.9k		24.4k	
	Words	183.3k	166.6k	189k	170k
	Vocabulary	10.5k	7.2k	10.9k	6.9k
Dev	Sentences	489		489	
	Words	5.8k	-	5.9k	-
	Vocabulary	1.2k	-	1.2k	-
	References	16		7	
Test	Sentences	500		500	
	Words	6.5k	-	3.2k	-
	Vocabulary	1.5k	-	926	-
	References	16		7	

Table A.2: *BTEC Arabic-English corpus. Basic statistics for the training, development and test data sets.*

		2006 Evaluation			
		Japanese	English	Italian	English
Train	Sentences	20k			
	Words	390.2k	324.8	155.4k	166.3k
	Vocabulary	10.6k	9.5k	10.2k	7.2k
Dev	Sentences	489		489	
	Words	6.8k	-	5.2 k	-
	Vocabulary	1.2k	-	1.2k	-
	References	16			
Test	Sentences	500		500	
	Words	7.3k	-	5.9k	-
	Vocabulary	1,3k	-	1.4k	-
	References	16			

Table A.3: *BTEC Japanese and Italian English corpus. Basic statistics for the training, development and test data sets.*

A.2   WMT data

The Workshop on Machine Translation (WMT) organizes an International Evaluation Campaign among European Languages. We participated several years and in different tasks as reported in Appendix § B.2. Additionally, given that the corpora were public, we used some versions to test several thesis approaches.

		2006 Evaluation		2007 Evaluation		2007 Evaluation	
		Spanish	English	Spanish	English	German	English
Train	Sentences	727,1k		1.30M		1.31M	
	Words	15.7M	15.2M	38.0M	35.6M	34.4M	36.0 M
	Vocabulary	108.7k	72.3k	138.9	119.0k	334.3k	136.7k
Dev	Sentences	500		500		1064	
	Words	15.2 k	14.8 k	15.3k	14.8k	26.9k	25.6k
	Vocabulary	3.6 k	3 k	3.5k	2.9k	6.6k	5.2k
	References	1					
Test	Sentences	3064		2000		2007	
	Words	91.9 k	85.2 k	62.6k	59.1k	50.9 k	49.6k
	Vocabulary	11.1 k	9.1 k	8.1k	6.4k	10.2k	7.5k
	References	1					

Table A.4: *2006 (Es2En) and 2007 (Es2En and De2En) WMT corpus. Basic statistics for the training, development and test data sets.*

Table A.4 reports the corpus statistics for 2006 EPPS Es2En task. For 2007 evaluations, training data is a concatenation of both EPPS and News Commentary corpora. There are about 35-40 million words of training data per language from the EPPS corpus and 1 million words from the News Commentary corpus. In the Es2En task, the test set belongs to the EPPS set (in-domain task) and in the De2En task, the test set belongs to the News Commentary set (out-domain task).

A.3   TC-STAR data

As explained in Chapter § 1.4, TC-STAR was an European project which aimed to advance research in all core technologies for Speech-to-Speech translation. In order to evaluate the project, periodic competitive evaluations were done. Therefore, the available corpora was constituted by the European Parliamentary Plenary Sessions (EPPS). As project participants, we participated in all the evaluations in the Spanish-English tasks as reported in Appendix § B.3.

All the experiments in this work are carried out over the *Final Text Edition* (FTE) version of the corpus. It mainly consists in text transcriptions of the Parliament speeches after edited in order to include punctuation, truecase and avoid different spontaneous speech phenomena. Three different corpora were released for the different TC-STAR evaluations.

		2nd Evaluation		3rd Evaluation	
		Spanish	English	Spanish	English
Train	Sentences	1.28M		1.35M	
	Words	32M	31M	39M	37M
	Vocabulary	159K	111K	147k	109k
Dev	Sentences	430	430	430	-
	Words	15.7k	16k	15.7k	-
	Vocabulary	3.2k	2.7k	3.2k	-
	References	3			
Test	Sentences	840	1094	892	-
	Words	22.7k	26.8k	29.1k	-
	Vocabulary	4k	3.9k	4.8k	-
	References	2			

Table A.5: *TC-STAR corpus of the three evaluations (Spanish-English task). Basic statistics for the training, development and test data sets.*

Table A.5 shows the corpus statistics for the 2nd and 3rd evaluations which were in 2006 and 2007, respectively. Experiments were carried out on an internal development and test set. The former was a random selection of the official one. The latter corresponded in each case (2006 and 2007) to the one year earlier official evaluation, i.e. for the 2nd evaluation, the 1st evaluation test set was used and for the 3rd evaluation, the 2nd evaluation test set was used.

B

Participation in International Evaluation Campaigns

This appendix reports the most relevant contributions to the MT International Evaluation Campaigns. The work presented here has to be understood in most cases as a team work and it complements this PhD scopes.

International evaluation campaigns are organized by different institutions, consortiums, conferences or workshops. These campaigns allow the sites, which participate in such an event comparing the state-of-the-art of their systems.

These evaluations support MT research and help advance the state of the art in MT technology. The goal of the international evaluation campaigns is to set up a framework for objectively comparing different MT systems. Through this, progress is to be promoted for MT technologies.

All submissions are usually evaluated automatically using different translation evaluation measures. In addition, sometimes the submitted translations are manually evaluated. Training data taken from different sources is explicitly specified. In general, the participating groups are not allowed to use any other data for developing their systems.

During this PhD we participated in several MT evaluations and we report a brief discussion about the results in the IWSLT (§ B.1), WMT (§ B.2), TC-STAR (§ B.3) and NIST (§ B.4). Section §B.5 gives the sites' acronyms.

B.1   International Workshop on Spoken Language Translation

Since 2004, the C-STAR [1] consortium organizes the International Workshop on Spoken Language Translation (IWLST) on a yearly basis. This workshop includes an evaluation campaign oriented towards speech translation. The objective is to provide a framework for the validation of existing evaluation methodologies concerning their applicability to the evaluation of spoken language translation technologies and open new directions on how to improve current methods. In order to achieve this goal and to support future evaluation research efforts, they released the supplied corpus and they obtained translation results after. The evaluation campaign was carried out using a multilingual speech corpus. It contains tourism-related sentences similar to those that are usually found in phrasebooks for tourists going abroad from the Basic Travel Expression Corpus (BTEC) § A.1. The test sets consist in translation of read speech in the travel domain.

B.1.1   2005 IWSLT Evaluation

In October 2005, the 2nd IWSLT took place in Pittsburgh [2]. Evaluation details can be found in (Doi et al., 2005).

We participated in the Chinese-to-English and Arabic-to-English tasks with a phrase-based system obtaining the best results in the latter.

The phrase-based model followed the log-linear maximum entropy framework, and it used as feature functions: the conditional and posterior probability, the source-to-target and target-to-source lexical probabilities, the word and phrase bonus and the distortion model (Crego et al., 2005b).

This phrase-based system has been improved in different ways. First, units were extracted from the concatenation of the union alignment and the source-to-target alignment. Second, it

---
[1]http://www.c-star.org/
[2]http://www.is.cs.cmu.edu/iwslt2005/

may not be efficient to consider a dictionary with all the phrases and the huge increase in computational and storage cost of including longer phrases does not provide a significant improve
in quality (Koehn et al., 2003) as the probability of reappearance of larger phrases decreases.
Therefore, we limited the maximum size of any given phrase. The length of a monolingual
phrase is defined as its number of words. The length of a phrase is the greatest of the lengths of
its monolingual phrases.

In our system, we considered two length limits. First we extracted all the phrases of length
X or less. Then, we also added phrases up to length Y (Y greater than X) if they could not be
generated by smaller phrases. Basically, we selected additional phrases with source words that
otherwise would be missed because of cross or long alignments (Costa-jussà and Fonollosa,
2005).

Table B.1 summarizes results for the Arabic-to-English and Chinese-to-English tasks.

Ar2En		Zh2En	
Site	BLEU	Site	BLEU
UPCph	**57.3 (1)**	ITC-irst	52.8
ITC-irst	56.2	RWTH	51.1
RWTH	54.7	EDINBG	46.5
IBM	53.8	**UPCph**	**45.2 (4)**
UPC	53.3	MIT/AF	45.0
EDINBG	51.1	UPC	44.4
NTT	44.6	CMU	44.4
CMU	40.9	IBM	44.0
USC-ISI	37.4	ATR-C3	39.4
		USC-ISI	33.2
		NTT	27.8

Table B.1: *Results of the 2005 IWSLT shared task. Our site rank is shown in parentheses.*

Details are reported in the following publication:

- Costa-jussà, M.R. and Fonollosa, J.A.R.
  Tuning a phrase-based statistical translation system for the IWSLT 2005 Chinese to English and Arabic to English tasks
  International Workshop on Spoken Language Translation (IWSLT) pp 185-190, Pittsburgh, 2005.

B.1.2   2006 IWSLT Evaluation

In November 2006, the 3rd IWSLT took place in Kyoto [3]. Evaluation details can be found in (Paul, 2006). Development and test sets were out-of-domain.

We participated in Chinese-to-English, Arabic-to-English, Italian-to-English and Japanese-to-English. Our system was a standard phrase-based system similar to the one reported in § 3.3.1 enriched with the statistical machine reordering technique (SMR-1best) (Costa-jussà et al., 2006). Additional local reordering (maxjumps § 5.1.1) was added in Chinese-to-English, Arabic-to-English and Japanese-to-English. TALPphr used the SMR reordering technique, which was expected to coherently improve the quality of translation in the evaluation set as it had in the internal set (see Table § 4.1). The high number of unknown words in the evaluation set may have caused a detriment of the SMR behavior at that time word classes for unknown words were still not disambiguated § 7.2.1.2.

We also participated with the TALPphr and TALPtup combination, the Ngram-based statistical machine translation system. The TALPcomb uses several n-gram language models, a word bonus and the IBM Model 1 for the whole sentence. The combination seems to obtain clear improvements in BLEU score but not in NIST, since the features that operate in the combination generally benefit shorter outputs. See TALPphr and TALPcomb participation in Table B.2.

One issue that reduced our system's competivity was the fact of not introducing the development set as training before doing the final translation. This was of great importance given that the development belonged to the SIGDAT corpus and the test as well, both were out-of-domain. Yet our results were state-of-the-art in all tasks (Paul, 2006).

Details are reported in the following publication:

> • Crego, J.M., de Gispert, A., Lambert, P., Khalilov, M., Costa-jussà, M.R., Mariño,J.B, Banchs, R. and Fonollosa, J.A.R.
> *TALP Ngram-Based Statistical Machine Translation for the IWSLT 2006*
> International Workshop on Spoken Language Translation (IWSLT) pp 116-122 Kyoto, November 2006.

---

[3]http://www.slt.atr.jp/IWSLT2006/

Zh2En		Ar2En		It2En		Jp2En	
Site	BLEU	Site	BLEU	Site	BLEU	Site	BLEU
RWTH	21.11	IBM	22.74	NiCT-ATR	29.89	RWTH	21.42
JHU WS06	18.63	TALPtup	21.36	TALPcom	28.37	NTT	19.84
MIT/AF	18.61	NiCT-ATR	21.17	TALPtup	28.18	NiCT-ATR	18.99
NTT	18.34	TALPcom	21.01	MIT/AF	27.98	MIT/AF	18.91
NiCT-ATR	17.75	NTT	20.71	ITC-irst	27.97	UKACMU	18.68
UKACMU	17.10	UKACMU	19.95	UW	27.87	ICT-irst	16.04
TALPcom	16.50	**TALPphr**	**19.08**	NTT	27.69	SLE	15.99
TALPtup	16.24	ITC-irst	17.23	**TALPphr**	**26.84**	HKUST	15.23
TALPphr	**15.99**	HKUST	14.77	DCU	25.98	Kyoto-U	14.18
Xiamen-U	15.79	DCU	14.50	UKACMU	23.88	TALPcom	13.90
ITC-irst	15.60	CLIPS	4.90	HKUST	23.74	TALPtup	13.70
HKUST	15.45	-	-	CLIPS	13.68	NAIST	13.11
ATT	12.26	-	-	-	-	**TALPphr**	**12.80**
NLPR	10.37	-	-	-	-	CLIPS	7.55

Table B.2: *Results of the 2006 IWSLT ASR output official evaluation shared task.*

B.1.3   2007 IWSLT Evaluation

In October 2007, the 4th IWSLT took place in Trento [4]. Evaluation details can be found in (Fordyce, 2007).

We participated in Chinese-to-English and Arabic-to-English with a system using POS-tag reordering rules §5.1.2 and introducing *six* additional feature functions: a target language model, a word bonus, two lexicon models, a target tagged language model and a source tagged (reordered) language model. Although all publicly available data was allowed, we only used the provided data to train the system.

Following a similar approach to that in (Habash and Sadat, 2006), we used the *MADA+TOKAN* system for disambiguation and tokenization of the Arabic training/development/test sets. For disambiguation only diacritic uni-gram statistics were employed. For tokenization we used the D3 scheme with -TAGBIES option. The D3 scheme splits the following set of clitics: w+, f+, b+, k+, l+, Al+ and pronominal clitics. The -TAGBIES option produces Bies POS tags on all taggable tokens. Chinese preprocessing included re-segmentation using ICTCLAS (Zhang et al., 2003) and POS tagging using the freely available Stanford Parser[5].

---

[4]http://iwslt07.itc.it/
[5]http://www-nlp.stanford.edu/software/lex-parser.shtml

Comparative results are summarized in Tables B.3 for Arabic and B.4 for Chinese, which includes manual evaluation scores. Regarding the human evaluation (%Best), it consists of the average number of times that a system was judged to be better than any other system (Callison-Burch et al., 2007). For each task, 300 sentences out of the 724 sentences in the evaluation set were randomly selected and presented to at least 3 evaluators. Since the ranking metric requires that each submission be compared to the other system outputs, each sentence may be presented multiple times but in the company of different sets of systems. Evaluators of each task and submission included 2 volunteers with experience in evaluating machine translation and 66 paid evaluators who were provided with a brief training in machine translation evaluation.

Site	Clean		ASR	
	%Better	BLEU	%Better	BLEU
DCU	45.1 (1)	0.4709	28.1	0.3942
UPC	**42.9 (2)**	**0.4804 (3)**	**31.8 (1)**	**0.4445 (1)**
UEKAE	36.4	0.4923 (1)	19.8	0.3679
UMD	36.0	0.4858 (2)	25.0	0.3908
UW	35.4	0.4161	26.9	0.4092
MIT	35.1	0.4553	31.4	0.4429
CMU	33.9	0.4463	25.5	0.3756
LIG	33.9	0.4135	24.2	0.3804
NTT	25.3	0.3403	25.5	0.3626
GREYC	21.7	0.3290	-	-
HKUST	13.1	0.1951	11.2	0.1420

Table B.3: *2007 IWSLT Arabic-to-English human (%Better) and automatic (BLEU) comparative results for the two tasks (Clean and ASR).*

Considering the Arabic-to-English pair, the UPC SMT system attained outstanding results, ranked in both cases (by human and automatic measures) as one of the best systems. Specially relevant is the performance achieved in the *ASR* task, where state-of-the-art results were obtained.

Notice that our system did not take multiple ASR output hypotheses into account but the single best. This gives additional relevance to the results achieved in the *ASR* task when compared to other systems.

The UPC SMT system falled in performance when considering the Chinese to English task. One of the reasons that can explain this situation is that our system seems to be less robust to noisy alignments (in special under scarce data availability) than standard phrase-based systems. The important reordering needs, the complexity of the Chinese vocabulary and the small data availability make the alignment process significantly more difficult in this translation task.

	Clean	
Site	%Better	BLEU
CASIA	37.6 (1)	0.3648 (5)
I2R	37.0 (2)	0.4077 (1)
ICT	34.8 (3)	0.3750 (2)
RWTH	32.4 (4)	0.3708 (4)
FBK	30.6 (5)	0.3472 (7)
CMU	30.6 (6)	0.3744 (3)
UPC	**28.3 (7)**	**0.2991 (11)**
XMU	28.1	0.2888
HKUST	25.5	0.3426 (8)
MIT	25.0	0.3631 (6)
NTT	24.6	0.2789
ATR	24.2	0.3133 (10)
UMD	23.6	0.3211
DCU	18.6	0.2737
NUDT	16.1	0.1934

Table B.4: *2007 IWSLT Chinese-to-English human (%Better) and automatic (BLEU) comparative results for the Clean task.*

Details are reported in the following publication:

- Lambert, P. and Costa-jussà, M.R. and Crego, J. M. and Khalilov, M. and Mariño, J. B. and Banchs, R. E. and R. Fonollosa, J. A. and Schwenk, H.
  The TALP Ngram-based SMT System for IWSLT 2007
  International Workshop on Spoken Language Translation (IWSLT07) pp 169-174, Trento, October 2007

B.2   Workshop on Statistical Machine Translation

Starting in 2005, the workshop featured a shared translation task. The organizers provided common test sets for translation between several European language pairs in both directions. The evaluation campaign was carried out using a multilingual corpus. It contains parliamentary transcriptions sentences from the European Plenary Parliamentary Speeches (EPPS). Through the years they have provided several new challenges as domain adaptation translation and human evaluation.

To lower the barrier of entrance to the competition, a complete baseline MT system, along with data, the following resources were provided: sentence-aligned training corpora, development and dev-test sets, language models trained for each language, an open source decoder for phrase-based SMT: Moses (Koehn et al., 2007), a training script to build models for Moses.

B.2.1   2005 WMT Evaluation

The first Workshop on Building and Using Parallel Texts: data driven MT and beyond was organized at the 43rd Annual Meeting of the Association for Computational Linguistics (ACL) and it also included a machine translation shared task reported in (Koehn and Monz, 2005).

We participated with an standard phrase-based system which included the same method for the phrase extraction as in § B.1.1. The phrase-based system included as feature functions: the conditional and posterior probability, the source-to-target and target-to-source lexicon probability and the word and phrase bonus. Decoding was monotonic. Table B.7 summarizes the results for the Spanish-to-English task.

Site	BLEU
UW	30.95
UPC	30.07
UPCm	**29.84 (3)**
NRC	29.08
RALI	28.49
UPCj	28.13
SAAR	26.69
CMU	26.14
UJI	21.65

Table B.5: *Results of the 2005 ACL shared task.*

Details are reported in the following publication:

- Costa-jussà, M.R. and Fonollosa, J.A.R.
  Improving Phrase-Based Statistical Translation by modifying phrase extraction and including new features
  Proc. of the ACL Workshop on Building and Using Parallel Texts (ACL'05/Wkshp), pp 133-136, Ann Arbor (Michigan), June 2005.

B.2.2   2006 WMT Evaluation

A second edition of Workshop on Machine Translation took place in Rochester in June 2006. Evaluation details can be found in (Koehn and Monz, 2006)

We participated with a standard phrase-based translation system enhanced with the RACBA algorithm (see details in Chapter § **??**). The baseline system is the same for all tasks and includes the same features functions as in § B.2.1. Additionally when the target language was English or Spanish we used a POS language model. The POStag analyzer was TNT and FreeLing respectively for each language.

Both Es2En and Fr2En tasks used the RABCA technique which improved the baseline system, see Table B.6. This technique was used when the source language belongs to the Romanic family. The length of the block is limited to 1 (i.e. it allows the swapping of single words). The main reason is that specific errors were solved in the tasks from a Romanic language to a Germanic language (as the common reorder of $Noun + Adjective$ that turns into $Adjective + Noun$).

Task	Baseline	+RABCA
Es2En	30.51	**31.41**
Fr2En	30.03	**30.33**

Table B.6: *Results evaluated on the test set. The bold results were the configurations submitted.*

In the opposite direction, En2Es and Fr2En used the reordering presented in (Crego et al., 2006a). A highly constrained reordered search performed by means of a set of reordering patterns (linguistically motivated rewrite patterns) which are used to extend the monotone search graph with additional arcs (§5.1.2).

Table B.7 summarizes the results for the tasks where we used reordering.

Details are reported in the following publication:

- Costa-jussà, M.R. and Fonollosa, J.A.R. *TALP Phrase-based Statistical Machine Translation System for European Language Pairs*
  HLT/NAACL 2006 Workshop on Statistical Machine Translation (WMT'06), pp 142-145, New York City, June 2006.

Es2En		En2Es		Fr2En		En2Fr	
Site	BLEU	Site	BLEU	Site	BLEU	Site	BLEU
LCC	31.46	**UPCm**	**31.06**	LCC	30.81	NTT	31.92
NTT	31.29	NTT	30.93	NTT	30.72	RALI	31.79
UTD	31.10	UTX	30.73	UTD	30.53	NRC	31.75
UPCjm	31.01	UPCjm	30.44	UPCjm	30.42	UPCjm	31.75
RALI	30.80	NRC	29.97	RALI	30.59	**UPCm**	**31.50**
NRCj	30.04	MS	29.76	**UPCm**	**30.53**	UTD	31.42
UPCm	**29.43**	RALI	29.38	CMU	30.18	SYSTRAN	25.07
EDINBG	29.01	EDINBG	28.49	NRC	29.62	-	-
UPCjg	28.03	UPCjg	27.46	UPV	24.10	-	-
UPC	23.91	UPV	23.17	SYSTRAN	21.44	-	-

Table B.7: *Results of the 2006 NAACL shared task.*

B.2.3   2007 WMT Evaluation

The shared task of the 2007 ACL Workshop on Statistical Machine Translation took place in Prague in June 2007. We participated in three language pairs: Spanish-English, French-English and German-English with translation tasks in both directions. Evaluation details of the shared task can be found in (Callison-Burch et al., 2007).

The shared task participants were provided with a common set of training and test data for all language pairs. The considered data was part of the European Parliament data set (Koehn, 2005), and included also News Commentary data, which was the surprise out-of-domain test set of the previous year (News Commentary corpus).

In addition to the EPPS test set, editorials from the Project Syndicate website[6] were collected and employed as out-of-domain test.

We participated with a standard Ngram-based system similar to the one reported in § 7.4.3.1 enhanced with an SMR-$Graph_R$ reordering approach. We participated in Spanish, French and German to English and the opposite directions. Our SMT approach was the BEST PERFORMING SYSTEM (both human and automatic evaluation) in the Spanish-English of the evaluation campaign (Callison-Burch et al., 2007). Our results specially stood out in the out-of-domain tasks, proving the capacity of generalization of the proposed reordering approach. Table B.8 summarizes the results.

The human evaluation was distributed across a number of people, including participants in

---

[6]http://www.project-syndicate.com/

Site	Adequacy	Fluency	BLEU
EPPS			
uedin	0.593 (1)	0.610 (1)	0.324 (1)
cmu-syn	0.552	0.568	0.323 (2)
upc	**0.587 (2)**	**0.604 (2)**	**0.322 (3)**
cmu-uka	0.557	0.564	0.320
nrc	0.477	0.489	0.313
upv	0.562	0.573	0.315
systran	0.525	0.566	0.290
saar	0.328	0.542	0.245
News			
upc	**0.566 (1)**	**0.543 (1)**	**0.346 (1)**
uedin	0.546	0.534	0.327
upv	0.435	0.459	0.283
cmu-uka	0.522	0.495	0.299
nrc	0.479	0.464	0.299
systran	0.525	0.503	0.259
saar	0.446	0.460	0.244

Table B.8: *2007 WMT Spanish-to-English human (Adequacy/Fluency) and automatic (BLEU) comparative results for the two tasks (EPPS and News).*

the shared task, interested volunteers, and a small number of paid annotators. More than one hundred people participated out of which at least seventy five employed at least one hour of effort to account for three hundred thirty hours of total effort.

UPC participated in all tasks except for the Czech-English with a system performing SMR reordering using a set of automatically extracted word classes(Costa-jussà and Fonollosa, 2006) and introducing *four* additional feature functions: a target language model, a word bonus and two lexicon models.

A Spanish morphology reduction was implemented in the preprocessing step, aiming at reducing the data sparseness problem due to the complex Spanish morphology. In particular, Spanish pronouns attached to the verb were separated, *i.e.* '*aplaudiéndola*' is transformed into '*aplaudiendo +la*'. And contractions like '*del*' were also separated into '*de el*'. GIZA++ alignments were performed after the preprocessing step.

Tables B.8 and B.9 detail respectively the Spanish-to-English and English-to-Spanish results. Human (*Adequacy* and *Fluency*) and automatic *BLEU*) measure are used for both translation tasks (*EPPS* and *News*).

Considering the Spanish-to-English results, the UPC SMT system obtains very competitive results, specially for the out-of-domain task (*News*), where the human and automatic measures reward the system with the best results.

Site	Adequacy	Fluency	BLEU
EPPS			
uedin	0.586 (1)	0.638 (1)	0.316 (1)
upc	**0.584 (2)**	**0.578 (4)**	**0.312 (2)**
cmu-uka	0.563	0.581 (3)	0.311
upv	0.573	0.587 (2)	0.304
nrc	0.546	0.548	0.299
systran	0.495	0.482	0.212
News			
ucb	0.449	0.414	0.331 (1)
upc	**0.510 (1/2)**	**0.488 (3)**	**0.328 (2)**
cmu-uka	0.510 (1/2)	0.492 (2)	0.327
uedin	0.429	0.419	0.322
nrc	0.408	0.392	0.311
upv	0.405	0.418	0.285
systran	0.501	0.507 (1)	0.281

Table B.9: *2007 WMT English-to-Spanish human (Adequacy/Fluency) and automatic (BLEU) comparative results for the two tasks (EPPS and News).*

In the case of the English to Spanish results, in spite of achieving also highly competitive results the UPC system slightly loses performance in the comparison against other systems. The preprocessing step reducing the Spanish vocabulary seems to help more the Spanish-to-English direction than the English-to-Spanish one.

Details are reported in the following publication:

- Costa-jussà, M.R., Crego, Josep M., Lambert, Patrik, Khalilov, Maxim, R. Fonollosa, José A., Mariño, José B. and Banchs, Rafael E.
  Ngram-Based Statistical Machine Translation Enhanced with Multiple Weighted Reordering Hypotheses
  Second Workshop of Statistical Machine Translation (WMT07) ACL pp 167-170, Prague, July 2007

B.3   TC-STAR Project

The European project TC-STAR (Technology and Corpora for Speech to Speech Translation) organized a first internal evaluation in 2005 (for members of the project, including UPC) and an open evaluation in 2006 and 2007.

To foster significant advances in translation technologies, periodic competitive evaluations were planned inside the project. A measure of success of the project was the involvement of external participants in the evaluation campaigns. Results were presented and discussed in a series of TC-STAR evaluation workshops. In addition to exchanging new knowledge through the evaluation workshops, these workshops were turned into public events that draw the attention of scientific community, industry, and in particular companies active in the area of technology transfer and services.

Spoken Language Translation evaluation was run in 3 translation directions: English-to-Spanish, Spanish-to-English and Mandarin-to-English.

English-to-Spanish and Spanish-to-English were run on recording transcriptions from the European Parliament Plenary Sessions (EPPS), while Chinese to English was run from recording transcriptions of Voice of America.

To study the effect of recognition errors and spontaneous speech phenomena, particularly for the EPPS task, three types of input to the translation system were studied and compared:

- **ASR**: the output of automatic speech recognizers, without using punctuation marks

- **verbatim**: the verbatim (i.e. correct) transcription of the spoken sentences including the phenomena of spoken language like false starts, ungrammatical sentences etc. (again without punctuation marks)

- **text**: the final text editions (FTE), which are the official transcriptions of the European Parliament and which do not include the effects of spoken language any more (here, punctuation marks were included)

In addition to these tasks, a complementary Spanish-to-English task was included in this evaluation for portability assessment in 2006 and 2007. This data consisted in transcriptions from Spanish Parliament, for which no parallel training was provided.

B.3.1   2005 TC-STAR Evaluation

In fall 2004, the TC-STAR EU-funded project organized its first evaluation campaign. UPC took part in the English-Spanish pair.

UPC participated with a monotonic version of the Ngram-based system with standard feature functions.

Results are summarized in Table B.11, where only FTE condition is shown. Results obtained by the UPC Ngram-based SMT system were very competitive with results from other sites [7].

Es2En		En2Es	
Site	BLEU	Site	BLEU
UPC	**53.3 (1)**	**UPC**	**46.2 (1)**
IBM	53.1	IBM	45.2
ITC-irst	47.5	RWTH	38.9
RWTH	46.1	UKA	37.6
UKA	40.5	UPV	34.1
UPV	32.7		

Table B.10: *2005 TC-STAR automatic scores obtained for the test data in EsEn FTE condition.*

B.3.2   2006 TC-STAR Evaluation

In February 2006, the TC-STAR EU-funded project organized its second evaluation campaign [8]. UPC took part in the same pair English-Spanish.

Results from B.11 are the official [9] from the 2nd evaluation of the TC-STAR project [10]. Our primary system was a standard Ngram-based translation system enhanced with the RABCA technique (see details § 6.3.1).

Details are reported in the following publication:

- Mariño,J.B., Banchs,R., Crego, J.M., de Gispert, A., Lambert, P., Fonollosa, J.A.R, Costa-jussà, M.R. and Khalilov, M.
  UPC's Bilingual N-gram Translation System
  TC-Star Workshop on Speech-to-Speech Translation TCSTAR'06 pp 43-48 Barcelona, June 2006.

---

[7]Note that a bug for RWTH Text results was reported after the evaluation
[8]http://www.elda.ord/tc-star-workshop/2006eval.htm
[9]for acronym's extension see B.5
[10]http://www.elda.ord/tc-star-workshop/2006eval.htm

Es2En		En2Es	
Site	BLEU	Site	BLEU
IBM	54.1	ITC-irst	49.8
RWTH	53.1	EDINBG	49.5
UW	52.8	UW	49.4
IRST	52.4	**UPC**	**48.2 (4)**
UPC	**52.3 (5)**	IBM	47.7
EDINBG	51.9	UKA	44.0
UKA	47.0	DFKI	36.3
SystrP	45.7	SystrP	36.3
DFKI	43.0		

Table B.11: *2006 TC-STAR automatic scores obtained for the test data in Es2En FTE condition.*

B.3.3 2007 TC-STAR Evaluation

In February 2007, the TC-STAR EU-funded project organized its third evaluation campaign. UPC took part in the same pair English-Spanish.

The Ngram-based SMT system presented to the evaluation was built from *unfold* translation units, and making use of POS-tag rules to account for reorderings. A set of 6 additional models were used: a target language model, a word bonus, a target tagged language model, a source (reordered) language model and two lexicon models computed on the basis of word-to-word translation probabilities.

We used as preprocessing step the same Spanish morphology reduction employed for the system built for the 2007 WMT evaluation, outlined in §B.2.3.

Notice that in the official results of the evaluation, multiple submissions for each participant and task are considered as well as a system combination '*ROVER*' that we have not introduced in this summary (we picked the best submission results of each participant when multiple were available).

Considering the Spanish-to-English results, the UPC system achieves very competitive results when compared to other participants. It is remarkable that our system is better ranked when the translation domain of the test data gets far from the domain employed to train the system. The system is better ranked not only when moving from *EPPS* to *Cortes*, but also from *FTE* to *Verbatim* and *ASR*. So far a single system is used for all tasks, which was built from data in the form of *FTE*, *Verbatim* and *ASR* tasks can be considered out-of-domain.

Es2En		En2Es	
Site	BLEU	Site	BLEU
IBM	0.5406 (1)	IBM	0.4754
ITC-irst	0.5240 (4)	ITC-irst	0.4981 (1)
RWTH	0.5310 (2)	RWTH	0.4944 (3)
UED	0.5187	UED	0.4950 (2)
UKA	0.4705	UKA	0.4404
UPC	**0.5230 (5)**	**UPC**	**0.4885 (4)**
DFKI	0.4304	DFKI	0.3632
UW	0.5261 (3)	UW	0.4850
SYSTRAN	0.4572	SYST	0.3629
LIMSI	-		

Table B.12: *2007 TC-STAR automatic scores obtained for the test data in Es2En FTE condition.*

Regarding English-to-Spanish results of Table B.12, the UPC system shows also a high level of competitivity, with scores close to those obtained by the best system. However, better ranking results are not observed for this translation direction when moving from *FTE* to *Verbatim* and *ASR*.

B.4   NIST Open Machine Translation Evaluation

Last but not least, with a large experience in automatic speech recognition benchmark tests, the National Institute of Standards and Technology (NIST), belonging to the Government of the United States, organizes yearly machine translation tests since the early 2000s. Aiming at a breakthrough in translation quality, these tests are usually unlimited in terms of data for training. The target language is English and sources include Arabic and Chinese [11].

The objective of the NIST Open Machine Translation (MT) evaluation series is to support research in, and help advance the state of the art of, technologies that translate text between human languages. Input may include all forms of text. The goal is for the output to be an adequate and fluent translation of the original.

The MT evaluation series started in 2001 as part of the DARPA TIDES program. In their current form, the evaluations are driven and coordinated by NIST as NIST Open MT. They provide an important contribution to the direction of research efforts and the calibration of

---

[11] http://www.nist.gov/speech/tests/mt/index.htm

technical capabilities in MT. The Open MT evaluations are intended to be of interest to all researchers working on the general problem of automatic translation between human languages. To this end, they are designed to be simple, to focus on core technology issues, to be fully supported, and to be accessible to all those wishing to participate.

NIST has carried out MT evaluation rounds for Mandarin to English translation. Additionally, the Arabic-to-English and Urdu-to-English translations have been added in the recent years.

B.4.1 2006 NIST Evaluation

The UPC SMT team participated by first time in the NIST Machine Translation evaluation on 2006.

The 2006 evaluation considered Arabic and Chinese the source languages under test, and English the target language. The text data consisted of newswire text documents, web-based newsgroup documents, human transcription of broadcast news, and human transcription of broadcast conversations. Performance was measured using BLEU. Human assessments were also taken into account on the evaluation, but only for the six best performing systems (in terms of BLEU).

Two evaluation data conditions were available for the participants: the (almost) *unlimited* data condition and the *large* data condition. The almost unlimited conditions has the single restriction of using data made available before February 2006. The large data conditions contemplates the use of data available from the LDC catalog.

UPC participated only on the large condition of both tasks (Chinese-to-English and Arabic-to-English). Unfortunately, we did not invest enough effort to prepare the evaluation before the test set release, what end up in a very poor data preprocessing of the huge amount of corpora available. The system was built performing *unfold* units, using heuristic constraints to allow for reordering (maximum distortion distance of 5 words and a limited number of 3 reordered words per sentence), and four additional models were employed: a target language model, a word bonus and two lexicon models.

Table B.13 shows the overall BLEU scores of both translation tasks. Results are sorted by the BLEU score of the NIST subset and reported separately for the GALE and the NIST subsets . Fully detailed results can be read in the NIST web site[12].

---

[12]http://www.nist.gov/speech/tests/mt/doc/mt06eval_official_results.html

	Ar2En			Zh2En	
Site	NIST	GALE	Site	NIST	GALE
google	0.4281 (1)	0.1826	isi	0.3393 (1)	0.1413
ibm	0.3954	0.1674	google	0.3316	0.1470 (1)
isi	0.3908	0.1714	lw	0.3278	0.1299
rwth	0.3906	0.1639	rwth	0.3022	0.1187
apptek	0.3874	0.1918 (1)	ict	0.2913	0.1185
lw	0.3741	0.1594	edin	0.2830	0.1199
bbn	0.3690	0.1461	bbn	0.2781	0.1165
ntt	0.3680	0.1533	nrc	0.2762	0.1194
itcirst	0.3466	0.1475	itcirst	0.2749	0.1194
cmu-uka	0.3369	0.1392	umd-jhu	0.2704	0.1140
umd-jhu	0.3333	0.1370	ntt	0.2595	0.1116
edin	0.3303	0.1305	nict	0.2449	0.1106
sakhr	0.3296	0.1648	cmu	0.2348	0.1135
nict	0.2930	0.1192	msr	0.2314	0.0972
qmul	0.2896	0.1345	qmul	0.2276	0.0943
lcc	0.2778	0.1129	hkust	0.2080	0.0984
upc	**0.2741 (17)**	**0.1149 (16)**	**upc**	**0.2071 (17)**	**0.0931 (17)**
columbia	0.2465	0.0960	upenn	0.1958	0.0923
ucb	0.1978	0.0732	iscas	0.1816	0.0860
auc	0.1531	0.0635	lcc	0.1814	0.0813
dcu	0.0947	0.0320	xmu	0.1580	0.0747
kcsl	0.0522	0.0176	lingua	0.1341	0.0663
			kcsl	0.0512	0.0199
			ksu	0.0401	0.0218

Table B.13: *2006 NIST Arabic-English and Chinese-English comparative results (in terms of BLEU) for the two subsets (NIST and GALE) of the large data condition.*

Unfortunately, both tasks results are far from the best system's results.

B.4.2   2008 NIST Evaluation

The 2008 NIST Open Machine Translation evaluation continues the ongoing series of evaluations of human language translation technology.

The UPC SMT team participated in Arabic-to-English and Urdu-to-English. We are reporting results for the latter as we were more actively involved.

Urdu-to-English was a highly restricted task. Only the LDC resource DVD designated specifically for this task was allowed for training. Resources that assist the core engine (such as segmenters, tokenizers, or taggers), as well as any rule development, were subject to the same restriction. These restrictions held for both the Urdu and the English side of the Urdu to English task.

Unfortunately, the data available was extremely noisy and it was aligned at the document level. We had no expert in Urdu so we were not able to properly clean the data. However, our system was ranked in the middle. Table B.14 shows the overall BLEU scores of the Ur2En translation task.

We participated with a standard Ngram-based system similar to the one reported in § 7.4.3.1 enhanced with an SMR-$Graph_{R+T}$ reordering approach. The alignment at the sentence level was done by using an adaptation of the BIA alignment (Lambert et al., 2007).

Ur2En	
Site	BLEU
google	22.81
bbn	20.28
IBM	20.26
isi-lw	19.83
UMD	18.29
MITLLAFRL	16.66
UPC	**16.14 (7)**
columbia	14.59
Edinburgh	14.56
NTT	0.1394
qmul	0.1338
CMU-XFER	10.16

Table B.14: *Results of the 2008 NIST Urdu-English task.*

B.5   Acronyms

apptek	Applications Technology Inc.	USA
ATR	ATR Spoken Language Communication Research Laboratory	Japan
auc	American University in Cairo	Egypt
bbn	BBN Technologies	USA
CASIA	Chinese Academy of Sciences Institute of Automation	China
CLIPS	Institut d'informatique el Mathématiques Appliqués de Grenoble	France
cmu	Carnegie Mellon University	USA
columbia	Columbia University	USA
DFKI	German Research Center for Artificial Intelligence	Germany
dcu	Dublin City University	Ireland
FBK	Fondazione Bruno Kesler	Italy
GREYC	University of Caen	France
hkust	Hong Kong University of Science and Technology	Hong Kong
ibm	IBM	USA
ict	Institute of Computing Technology Chinese Academy of Sciences	China
iscas	Institute of Software Chinese Academy of Sciences	China
I2R	Institute for Infocomm Research	Singapore
isi	Information Sciences Institute	USA
itcirst	ITC-irst	Italy
ksu	Kansas State University	USA
kcsl	KCSL Inc.	Canada
kyoto-U	Kyoto University	Japan
lw	Language Weaver	USA
lcc	Language Computer	USA
LIG	University J. Fourier	France
LIMSI	Lab. d'Informatique pour la Mécanique et les Science de l'Ingeni.	France
lingua	Lingua Technologies Inc.	Canada
MISTRAL	University of Montreal Canada and University of Avignon	France
MIT/AF	Massachesetts Institute of Technology and Air Force	USA
msr	Microsoft Research	USA
naist	Nara Institute of Science and Technology	Japan

nict	National Institute of Information and Communications Technology	Japan
nlpr	National Laboratory of Pattern Recognition	China
ntt	NTT Communication Science Laboratories	Japan
nrc	National Research Council Canada	Canada
NUDT	National University of Defense Technology	China
qmul	Queen Mary University of London	England
RALI	Dep. d'Informatique et de Recherche Operationnelle in Montreal	Canada
RWTH	Rheinish-Westphalian Technical University	Germany
saar	Saarland University	Germany
sakhr	Sakhr Software Co.	USA
sri	SRI International	USA
stanford	Stanford University	USA
SYST	SYSTRAN Language Translation Software	France
ucb	University of California Berkeley	USA
uedin	University of Edinburgh	Scotland
UJI	Universitat Jaume I	Spain
upenn	University of Pennsylvania	USA
upc-lsi	Universitat Politechnica de Catalunya, LSI	Spain
upc-talp	Universitat Politechnica de Catalunya, TALP	Spain
upv	Technical University of Valencia	Spain
usc	University of Southern California	USA
UTX	University of Texas	US
xmu	Xiamen University	China
iai	Institute of Artificial Intelligence	China
uka	University of Karlsruhe	Germany
umd	University of Maryland	USA
JHU	Johns Hopkins University	USA
UW	University of Washington	USA

Publications by the Author

The work developed in this PhD has lead to the following publications.

- *State-of-the-art Word Reordering Approaches in Statistical Machine Translation*
  Costa-jussà, M.R. and Fonollosa, J.A.R.
  IEICE Transactions on Information and Systems vol 92, num 11, pp 2179-2185, 2009

- *Phrase and Ngram-based Statistical Machine Translation System Combination*
  Costa-jussà, M.R. and Fonollosa, J.A.R.
  Applied Artificial Intelligence: An International Journal vol 23, num 7, pp 694-711, 2009

- *An Ngram-based reordering model*
  Costa-jussà, M.R. and Fonollosa, J.A.R.
  Computer Speech and Language vol 23, num 3, issn 0885-2308, pp 362-375, 2009

- *Using Reordering in Statistical Machine Translation based on Alignment Block Classification*
  Costa-jussà, M.R. and Fonollosa, J.A.R. and Monte, E.
  Conference on Language Resources and Evaluation (LREC'08) Marraqueix, May 2008

- Lambert, P. and Costa-jussà, M.R. and Crego, J. M. and Khalilov, M. and Mariño, J. B. and Banchs, R. E. and R. Fonollosa, J. A. and Schwenk, H.
  The TALP Ngram-based SMT System for IWSLT 2007
  International Workshop on Spoken Language Translation (IWSLT07) pp 169-174, Trento, October 2007

- Schwenk, H., Costa-jussà, M.R. and Fonollosa, J.A.R.
  Smooth Bilingual Translation
  Empirical Methods in Natural Language Processing (EMNLP) pp 430-438, Prague, June July 2007.

- Costa-jussà, M.R. and Fonollosa, J.A.R.
  Analysis of Statistical and Morphological Classes to Generate Weigthed Reordering Hypotheses on a Statistical Machine Translation System
  Second Workshop of Statistical Machine Translation (WMT07) ACL pp 171-176, Prague, July 2007

- Costa-jussà, M.R., Crego, J.M., de Gispert, A., Lambert, P., Khalilov, M., Fonollosa, J.A.R., Mariño,J-B,and Banchs, R.
  Ngram-Based Statistical Machine Translation Enhanced with Multiple Weighted Reordering Hypotheses
  Second Workshop of Statistical Machine Translation (WMT07) ACL pp 167-170, Prague, July 2007

- Costa-jussà, M.R., Crego, J.M., Vilar, D., Fonollosa, J.A.R., Mariño, J.B. and Ney, H..
  Analysis and System Combination of Phrase- and Ngram-based Statistical Machine Translation Systems
  HLT-NAACL Conference pp 137-140, Rochester, May 2007

- Mariño,J.B., Banchs,R., Crego, J.M., de Gispert, A., Lambert, P., Fonollosa, J.A.R and Costa-jussà, M.R.
  Ngram-based Translation System
  Computational Linguistics, Volume 4, Number 32, pp 527-549, December 2006.

- Lambert, P., Giménez, J., Costa-jussà, M.R., Amigó, E., Banchs, R.E., Màrquez, LL. and Fonollosa, J.A.R.
  Machine Translation System Development Based on Human Likeness
  IEEE Workshop on Spoken Language Translation (SLT), Aruba, December 2006.

- Schwenk, H., Costa-jussà, M.R. and Fonollosa, J.A.R.
  Continuous Space Language Models for the IWSLT 2006 task
  International Workshop on Spoken Language Translation (IWSLT) pp 166-173, Kyoto,
  November 2006.

- Costa-jussà, M.R., Crego, J.M., de Gispert, A., Lambert, P., Khalilov, M., Fonollosa,
  J.A.R., Mariño,J-B,and Banchs, R.
  *TALP Phrase-based Statistical Machine Translation and TALP system combination the
  IWSLT 2006*
  International Workshop on Spoken Language Translation (IWSLT) pp 123-129, Kyoto,
  November 2006.

- Crego, J.M., de Gispert, A., Lambert, P., Khalilov, M., Costa-jussà, M.R., Mariño,J.B,
  Banchs, R. and Fonollosa, J.A.R.
  TALP Ngram-Based Statistical Machine Translation for the IWSLT 2006
  International Workshop on Spoken Language Translation (IWSLT) pp 116-122, Kyoto,
  November 2006.

- Costa-jussà, M.R. and Fonollosa, J.A.R.
  Sistema de Reordenamiento Estadístico
  Procesamiento del Lenguaje Natural (SEPLN06), pp 249-256, Zaragoza, Septiembre
  2006.

- Costa-jussà, M.R. and Fonollosa, J.A.R.
  Statistical Machine Reordering
  Empirical Methods in Natural Language Processing (EMNLP06) pp 70-77, Sydney July
  2006.

- Mariño,J.B., Banchs,R., Crego, J.M., de Gispert, A., Lambert, P., Fonollosa, J.A.R, Costa-
  jussà, M.R. and Khalilov, M.
  UPC's Bilingual N-gram Translation System
  TC-Star Workshop on Speech-to-Speech Translation TCSTAR'06 pp 43-48, Barcelona,
  June 2006.

- Costa-jussà, M.R. and Fonollosa, J.A.R.
  TALP Phrase-based Statistical Machine Translation System for European Language Pairs
  HLT/NAACL 2006 Workshop on Statistical Machine Translation (WMT'06), pp 142-
  145, New York City, June 2006.

- Crego, J.M., de Gispert, A., Lambert, P., Costa-jussà, M.R., Khalilov, M., Banchs, R., Mariño, J.B. and Fonollosa, J.A.R.
  N-gram-based SMT System Enhanced with Reordering Patterns
  HLT/NAACL 2006 Workshop on Statistical Machine Translation pp 162-165, New York City, June 2006.

- Crego,J.M., Costa-jussà,M.R., Mariño,J.B. and Fonollosa, J.A.R.
  Ngram-based versus Phrase-based Statistical Machine Translation
  International Workshop on Spoken Language Translation (IWSLT) pp 177-184, Pittsburgh, 2005.

- Costa-jussà, M.R. and Fonollosa, J.A.R.
  Tuning a phrase-based statistical translation system for the IWSLT 2005 Chinese to English and Arabic to English tasks
  International Workshop on Spoken Language Translation (IWSLT) pp 185-190, Pittsburgh, 2005.

- Mariño, J.B., Banchs, R.E., Crego, J.M., de Gispert, A., Lambert, P., Fonollosa, J.A.R. and Ruiz,M.
  Bilingual N-gram Statistical Machine Translation
  Proc. of the 10th Machine Translation Summit (MTsummit'05), pp 275-82, Pukhet (Thailand), Sep 2005.

- Costa-jussà, M.R. and Fonollosa, J.A.R.
  Técnicas mejoradas para la traducción basada en frases
  Procesamiento del Lenguaje Natural, núm. 35 (SEPLN'05) pp 351-357, Granada (Spain), Sep 2005.

- Mariño, J.B., Banchs, R., Crego, J.M., de Gispert, A., Lambert, P., Fonollosa, J.A.R. and Costa-jussà, M.R.
  Modelo estocástico de traducción basado en N-gramas de tuplas bilingües y combinación log-lineal de características
  Procesamiento del Lenguaje Natural, núm. 35 (SEPLN'05), pp 69-76, Granada (Spain), Sep 2005.

- Costa-jussà, M.R. and Fonollosa, J.A.R.
  Improving Phrase-Based Statistical Translation by modifying phrase extraction and including new features

Proc. of the ACL Workshop on Building and Using Parallel Texts (ACL'05/Wkshp), pp 133-136, Ann Arbor (Michigan), June 2005.

- 2004 Ruiz Costa-jussà,M., Gauvain, J.L, y Galibert, O.
Normalización de textos y selección del vocabulario para estimar el modelo de lenguaje de un sistema de transcripción de noticias
III Jornadas de Tecnologías del Habla (IIIJTH), Valencia, October 2004.

Bibliography

Y. Al-Onaizan, J. Curin, M. Jahr, K. Knight, J. Lafferty, D. Melamed, F.J. Och, D. Purdy, N.A. Smith, and D. Yarowsky. Statistical machine translation: Final report. Technical report, Johns Hopkins University Summer Workshop, Baltimore, MD, USA, 1999.

E. Amigó, J. Giménez, J. Gonzalo, and L. Màrquez. MT evaluation: Human-like vs. human acceptable. In *Proc. of the COLING/ACL 2006 Main Conf. Poster Sessions*, pages 17–24, Sydney, Australia, July 2006.

D. Arnold and L. Balkan. Machine translation: an introductory guide. *Computational Linguistics*, 21(4):577–578, 1995.

J. Atserias, B. Casas, E. Comelles, M. González, L. Padró, and M. Padró. Freeling 1.3: Syntactic and semantic services in an open-source nlp library. In *5th Int. Conf. on Language Resource and Evaluation (LREC)*, pages 184–187, 2006.

S. Bangalore and A. Joshi. Supertagging: An approach to almost parsing. *Computational Linguistics*, 25(2):237–265, 1999.

S. Bangalore and G. Riccardi. Stochastic finite-state models for spoken language machine translation. *Proc. Workshop on Embedded Machine Translation Systems*, pages 52–59, April 2000a.

S. Bangalore and G. Riccardi. Finite-state models for lexical reordering in spoken language translation. In *Proc. of the 6th Int. Conf. on Spoken Language Processing, ICSLP'02*, volume 4, pages 422–425, Beijing, October 2000b.

S. Bangalore, G. Bordel, and G. Riccardi. Computing consensus translation from multiple machine translation systems. In *IEEE Workshop on Automatic Speech Recognition and Understanding*, pages 351–354, Madonna di Campiglio, Italy, 2001.

A. Berger, S. Della Pietra, and V. Della Pietra. A maximum entropy approach to natural language processing. *Computational Linguistics*, 22(1):39–72, March 1996.

F. Vanden Berghen and H. Bersini. CONDOR, a new parallel, constrained extension of Powell's UOBYQA algorithm: Experimental results and comparison with the DFO algorithm. *Journal of Computational and Applied Mathematics*, 181:157–175, 2005.

N. Bertoldi. Minimum error training (updates). Technical report, Slides of the JHU Summer Workshop, 2006.

N. Bertoldi, R. Cattoni, M. Cettolo, B. Chen, and M. Federico. ITC-irst at the 2006 TC-STAR SLT evaluation campaign. In *TC-STAR Workshop on Speech-to-Speech Translation*, pages 19–24, Barcelona, Spain, June 2006.

B. Bonnie. Machine translation divergences: a formal description and proposed solution. *Computational Linguistics*, 20(4):597–633, 1994.

T. Brants. TnT – A statistical part-of-speech tagger. In *Proc. of the Sixth Applied Natural Language Processing (ANLP-2000)*, pages 224–231, Seattle, WA, 2000.

C. Brew and H. Thompson. Automatic evaluation of computer generated text: A progress report on the texte- val project. In *In Human Language Technology: Proceedings of the Workshop (ARPA/ISTO)*, pages 108–113, 1994.

P. Brown, J. Cocke, S. Della Pietra, V. Della Pietra, F. Jelinek, J.D. Lafferty, R. Mercer, and P.S. Roossin. A statistical approach to machine translation. *Computational Linguistics*, 16(2): 79–85, 1990.

P. Brown, S. Della Pietra, V. Della Pietra, and R. Mercer. The mathematics of statistical machine translation. *Computational Linguistics*, 19(2):263–311, 1993.

C. Callison-Burch, D. Talbot, and M. Osborne. Statistical machine translation with word- and sentence-aligned parallel corpora. In *Proc. of the 42th Annual Meeting of the Association for Computational Linguistics*, pages 175–182, Barcelona, Spain, July 2004.

C. Callison-Burch, M. Osborne, and P. Koehn. Re-evaluating the role of BLEU in machine translation research. In *Proc. of the 11th Conf. of the European Chapter of the Association for Computational Linguistics*, pages 249–256, Trento, Italy, March 2006.

C. Callison-Burch, C. Fordyce, P. Koehn, C. Monz, and J. Schroeder. (meta-) evaluation of machine translation. In *Annual Meeting of the Association for Computational Linguistics:*

Proc. of the Second Workshop on Statistical Machine Translation, pages 136–158, Prague, June 2007.

X. Carreras, I. Chao, L. Padró, and M. Padró. Freeling: An open-source suite of language analyzers. In *4th Int. Conf. on Language Resources and Evaluation, LREC'04*, Lisboa, Portugal, May 2004.

F. Casacuberta. Finite-state transducers for speech-input translation. In *IEEE Automatic Speech Recognition and Understanding Workshop, ASRU*, pages 375–380, Trento, 2001.

F. Casacuberta and E. Vidal. Machine translation with inferred stochastic finite-state transducers. *Computational Linguistics*, 30(2):205–225, 2004.

Pi-Chuan Chang and Kristina Toutanova. A discriminative syntactic word order model for machine translation. In *Proc. of the 45th Annual Meeting of the Association for Computational Linguistics*, pages 9–16, Prague, Czech Republic, June 2007.

E. Charniak, K. Knight, and K. Yamada. Syntax-based language models for machine translation. In *MT Summit*, New Orleans, 2003.

B. Chen, R. Cattoni, N. Bertoldi, M. Cettolo, and M. Federico. The ITC-irst statistical machine translation system for IWSLT-2005. In *Proc.of the Int. Workshop on Spoken Language Translation, IWSLT'05*, pages 98–104, Pittsburgh, October 2005.

S. F. Chen and J. T. Goodman. An empirical study of smoothing techniques for language modeling. *Computer Speech and Language*, 13(4):359–394, 1999.

S. F. Chen and J. T. Goodman. An empirical study of smoothing techniques for language modeling. Technical report, Harvard University, 1998.

D. Chiang. A hierarchical phrase-based model for statistical machine translation. In *Proc. of the 43rd Annual Meeting of the Association for Computational Linguistics (ACL'05)*, pages 263–270, Ann Arbor, Michigan, June 2005.

M. Collins. *Head-driven Statistical Models for Natural Language Parsing*. PhD Thesis, University of Pennsylvania, 1999.

M. Collins, P. Koehn, and I. Kucerová. Clause restructuring for statistical machine translation. In *Proc. of the 43th Annual Meeting of the Association for Computational Linguistics*, pages 531 – 540, Michigan, 2005.

M. R. Costa-jussà and J.A.R. Fonollosa. Improving the phrase-based statistical translation by modifying phrase extraction and including new features. In *Proceedings of the ACL Workshop on Building and Using Parallel Texts: Data-Driven Machine Translation and Beyond*, pages 133–136, Ann Arbor, MI, 2005.

M. R. Costa-jussà, J. M. Crego, A. de Gispert, P. Lambert, M. Khalilov, J. B. Mariño, J. A. R. Fonollosa, and R. Banchs. TALP phrase-based statistical translation system for European language pairs. In *Human Language Technology Conf. (HLT-NAACL'06): Proc. of the Workshop on Statistical Machine Translation*, pages 142–145, New York City, June 2006.

M.R. Costa-jussà and J.A.R. Fonollosa. Statistical machine reordering. In *Proc. of the Conf. on Empirical Methods in Natural Language Processing, EMNLP'06*, pages 71–77, Sydney, July 2006.

M.R. Costa-jussà, J.M. Crego, A. de Gispert, P. Lambert, M. Khalilov, J.A.R. Fonollosa, J-B. Mariño, and R. Banchs. TALP phrase-based statistical machine translation and TALP system combination the IWSLT 2006. In *Proc.of the Int. Workshop on Spoken Language Translation, IWSLT'06*, Kyoto, November 2006.

J. M. Crego. *Architecture and Modeling for N-gram-based Statistical Machine Translation*. PhD thesis, Department of Signal Theory and Communications, Universitat Politècnica de Catalunya (UPC), 2008.

J. M. Crego, M. R. Costa-jussà, J. Mariño, and J. A. Fonollosa. Ngram-based versus phrase-based statistical machine translation. *Proc.of the Int. Workshop on Spoken Language Translation, IWSLT'05*, pages 177–184, October 2005a.

J. M. Crego, A. de Gispert, P. Lambert, M. R. Costa-jussà, Maxim Khalilov, R. Banchs, José B. Mariño, and J. A. R. Fonollosa. N-gram-based SMT system enhanced with reordering patterns. In *Human Language Technology Conf. (HLT-NAACL'06): Proc. of the Workshop on Statistical Machine Translation*, pages 162–165, New York City, June 2006a.

J. M. Crego, A. de Gispert, P. Lambert, M. Khalilov, M.R. Costa-jussà, J. Mariño, R. Banchs, and J. A. Fonollosa. The TALP Ngram-based system for the IWSLT2006. In *Proc.of the Int. Workshop on Spoken Language Translation, IWSLT'06*, Kyoto, November 2006b.

J.M. Crego and J.B. Mariño. Improving SMT by coupling reordering and decoding. *Machine Translation*, 20(3):199–215, 2007.

J.M. Crego, J. Mariño, and A. de Gispert. An Ngram-based statistical machine translation decoder. In *Proc. of the 9th Int. Conf. on Spoken Language Processing, ICSLP'05*, pages 3185–3188, Lisboa, April 2005b.

J.M. Crego, A. de Gispert, P. Lambert, M.R. Costa-jussà, M. Khalilov, R. Banchs, J.B. Mariño, and J.A.R. Fonollosa. N-gram-based SMT system enhanced with reordering patterns. In *Human Language Technology Conf. (HLT-NAACL'06): Proc. of the Workshop on Statistical Machine Translation*, pages 162–165, New York City, June 2006c.

A. de Gispert. *Introducing Linguistic Knowledge into Statistical Machine Translation*. PhD Thesis in Computational Linguistics, Dep. de Teoria del Senyal i Comunicacions, Universitat Politècnica de Catalunya (UPC), 2006.

A. de Gispert and J.B. Mariño. Using x-grams for speech-to-speech translation. In *Proc. of the 7th Int. Conf. on Spoken Language Processing, ICSLP'02*, pages 1885–1888, Denver, 2002.

G. Doddington. Automatic evaluation of machine translation quality using n-gram co-occurrence statistics. In *Proc. of the Human Language Technology Conf., HLT-NAACL'02*, pages 138 – 145, San Diego, 2002.

T. Doi, Y. Hwang, K. Imamura, H. Okuma, and E. Sumita. Nobody is perfect: Atr s hybrid approach to spoken language translation. In *Proc.of the Int. Workshop on Spoken Language Translation, IWSLT'04*, pages 55–62, Pittsburgh, PA, USA, 2005.

B.J. Dorr. Machine translation: a view from the lexicon. *Computational Linguistics*, 20(4): 670–676, 1994.

G. Fiscus. A post-processing system to yield reduced word error rates: Recognizer output voting error reduction (ROVER). In *IEEE Workshop on Automatic Speech Recognition and Understanding*, 1997.

C. Fordyce. Overview of the IWSLT 2007 Evaluation Campaign. In *Proc.of the Int. Workshop on Spoken Language Translation, IWSLT'07*, pages 1–12, Trento, Italy, 2007.

G. Foster, R. Kuhn, and H. Johnson. Phrasetable smoothing for statistical machine translation. In *Proc. of the Conf. on Empirical Methods in Natural Language Processing, EMNLP'06*, pages 53–61, Sydney, Australia, 2006.

A. Fraser and D. Marcu. Measuring word alignment quality for statistical machine translation. Technical report, ISI/University of Southern California, California, 2006.

R. Frederking and S. Nirenburg. Three heads are better than one. In *4th Conference on Applied Natural Language Processing*, Stuttgart, German, 1994.

M. Galley and M. Hopkins. What's in a translation rule? In *Proc. of the Human Language Technology Conf., HLT-NAACL'04*, pages 273–280, Boston, MA, May 2004.

J. Giménez and E. Amigó. IQMT: A framework for automatic machine translation evaluation. *5th Int. Conf. on Language Resources and Evaluation, LREC'06*, pages 22–28, May 2006.

J. Giménez and L. Màrquez. Context-aware discriminative phrase selection for statistical machine translation. In *Annual Meeting of the Association for Computational Linguistics: Proc. of the Second Workshop on Statistical Machine Translation (WMT)*, pages 159–166, Prague, Czech Republic, June 2007.

A. O. González, G. Boleda, M. Melero, and T. Badia. Traducción automática estadística basada en n-gramas. *Procesamiento del Lenguaje Natural, SELPN*, 35:69–76, 2005.

N. Habash. Syntactic preprocessing for statistical machine translation. *Proc. of the MT-Summit XI*, September 2007.

N. Habash and O. Rambow. Arabic tokenization, part-of-speech tagging and morphological disambiguation in one fell swoop. In *Proc. of the 43th Annual Meeting of the Association for Computational Linguistics*, pages 573–580, Ann Arbor, MI, June 2005.

N. Habash and F. Sadat. Arabic preprocessing schemes for statistical machine translation. In *Proceedings of the Human Language Technology Conference of the NAACL, Companion Volume: Short Papers*, pages 49–52, New York City, USA, June 2006.

S. Hasan and H. Ney. Clustered language models based on regular expressions for statistical machine translation. *10th Annual Conference of The European Association for Machine Translation (EAMT)*, pages 119–125, 2005.

H. Hassan, M. Hearne, A. Way, and K. Sima'an. Syntactic phrase-based statistical machine translation. *1st IEEE/ACL Workshop on Spoken Language Technology*, December 2006.

H. Hassan, K. Sima'an, and A. Way. Supertagged phrase-based statistical machine translation. In *Proc. of the 45th Annual Meeting of the Association for Computational Linguistics*, pages 288–295, Prague, Czech Republic, June 2007.

S. Hewavitharana, B. Zhao, A. Silja Hildebrand, M. Eck, C. Hori, S. Vogel, and A. Waibel. The CMU statistical machine translation system for IWSLT2005. In *Proc.of the Int. Workshop on Spoken Language Translation, IWSLT'05*, 2005.

W. Hutchins. Machine translation: past, present and future. *Ellis Horwood*, 1986.

W.J. Hutchins and H.L. Somers. *An introduction to machine translation*. Academic Press, London, UK, 1992.

S. Jayaraman and A. Lavie. Multi-engline machine translation guided by explicit word matching. In *10th Conference of the European Association for Machine Translation*, pages 143–152, Budapest,Hungary, 2005.

F. Jelinek. *Statistical Methods for Speech Recognition (Language, Speech, and Communication)*. The MIT Press, January 1998.

S. Kanthak, D. Vilar, E. Matusov, R. Zens, and H. Ney. Novel reordering approaches in phrase-based statistical machine translation. In *Annual Meeting of the Association for Computational Linguistics: Proc. of the ACL Workshop on Building and Using Parallel Texts: Data-Driven Machine Translation and Beyond (WMT)*, pages 167–174, Ann Arbor, MI, June 2005.

M. Kay, J. Gawron, and P. Norvig. Verbmobil: a translation system for face-to-face dialog. *Linguistic Issues in Language Technology, CSLI*, 1992.

M. King, A. Popescu-Belis, and E. Hovy. FEMTI: creating and using a framework for MT evaluation. In *Proc. of the 5th Conf. of the Association for Machine Translation in the Americas, AMTA'03*, pages 224–231, 2003.

K. Kirchhoff and M. Yang. Improved language modeling for statistical machine translation. In *ACL'05 workshop on Building and Using Parallel Text*, pages 125–128, 2005.

R. Kneser and H. Ney. Improved backing-off for m-gram language modeling. In *Proc. of the ICASSP Conference*, volume 1, pages 181–184, 1995.

K. Knight. Decoding complexity in word-replacement translation models. *Computational Linguistics*, 25(4), December 1999a.

K. Knight. A statistical machine translation tutorial workbook. http://www.clsp.jhu.edu/ws99/projects/mt/wkbk.rtf, April 1999b.

K. Knight and Y. Al-Onaizan. Translation with finite-state devices. In *Proc. of the 4th Conf. of the Association for Machine Translation in the Americas, AMTA'02*, pages 421–437, Langhorne, December 1998.

K. Koehn and K. Knight. Empirical methods for compound splitting. *Proc. of the 10th Conf. of the European Chapter of the Association for Computational Linguistics*, pages 347–354, April 2003a.

P. Koehn. Europarl: A parallel corpus for statistical machine translation. *MT Summit*, pages 79–86, September 2005.

P. Koehn and K. Knight. Feature-rich statistical translation of noun phrases. In *Proc. of the 41th Annual Meeting of the Association for Computational Linguistics*, pages 311–318, 2003b.

P. Koehn and C. Monz. Shared task: Statistical machine translation between european languages. In *Annual Meeting of the Association for Computational Linguistics: Proc. of the Workshop on Statistical Machine Translation (WMT)*, pages 119–124, Michigan, June 2005.

P. Koehn and C. Monz. Manual and automatic evaluation of machine translation between european languages. In *Proc. of the Workshop on Statistical Machine Translation*, pages 102–121, New York City, June 2006.

P. Koehn, F.J. Och, and D. Marcu. Statistical phrase-based translation. In *Proc. of the Human Language Technology Conf., HLT-NAACL'03*, pages 48–54, Edmonton, Canada, May 2003.

P. Koehn, A. Amittai, A. Birch, C. Callison-Burch, M. Osborne, D. Talbot, and M. White. Edinburgh system description for the 2005 iwslt speech translation evaluation. In *Proc. of International Workshop on Spoken Languages Translation*, Pittsburgh, October 2005.

P. Koehn, H. Hoang, A. Birch, C. Callison-Burch, M. Federico, N. Bertoldi, B. Cowan, W. Shen, C. Moran, R. Zens, C. Dyer, O. Bojar, A. Constantin, and E. Herbst. Moses: Open source toolkit for statistical machine translation. In *Proc. of the 45th Annual Meeting of the Association for Computational Linguistics*, pages 177–180, Prague, Czech Republic, June 2007.

S. Kumar and W. Byrne. Minimum bayes-risk decoding for statistical machine translation. In *Proc. of the Human Language Technology Conf., HLT-NAACL'04*, pages 169–176, Boston, Massachusetts, USA, May 2004.

P. Lambert. *Exploiting Lexical Information and Discriminative Alignment Training in Statistical Machine Translation*. PhD thesis, Software Department, Universitat Politècnica de Catalunya (UPC), 2008.

P. Lambert and R. E. Banchs. Tuning Machine Translation Parameters with SPSA. In *Proc. of the International Workshop on Spoken Language Translation*, pages 190–196, Kyoto, Japan, 2006.

P. Lambert, A. de Gispert, R. Banchs, and J. Mariño. Guidelines for word alignment and manual alignment. *Language Resources and Evaluation*, 39(4):267–285, 2006.

P. Lambert, R. E. Banches, and J.M. Crego. Discriminative alignment training without annotated data for machine translation. In *Proc. of the Human Language Technology Conf., HLT-NAACL'07*, pages 199–215, Rochester,USA, 2007.

P. Langlais and F. Gotti. Phrase-based SMT with shallow tree-phrases. *Proceedings of the Workshop on Statistical Machine Translation*, pages 39–46, June 2006.

A. Lavie and A. Agarwal. METEOR: An automatic metric for MT evaluation with high levels of correlation with human judgments. In *Annual Meeting of the Association for Computational Linguistics: Proc. of the Second Workshop on Statistical Machine Translation (WMT)*, pages 228–231, Prague, Czech Republic, June 2007.

Y.-S. Lee, Y. Al-Onaizan, K. Papineni, and S. Roukos. IBM spoken language translation system. In *TC-STAR Workshop on Speech-to-Speech Translation*, pages 13–18, Barcelona, Spain, June 2006.

D. Marcu, Wong. W, A. Echihabi, and K. Knight. SPMT: Statistical machine translation with syntactified target language phrases. In *Proc. of the Conf. on Empirical Methods in Natural Language Processing, EMNLP'06*, pages 44–52, Sydney, Australia, July 2006.

J.B. Mariño, R.E. Banchs, J.M. Crego, A. de Gispert, P. Lambert, J.A.R. Fonollosa, and M.R. Costa-jussà. N-gram based machine translation. *Computational Linguistics*, 32(4):527–549, December 2006.

Y. Matsumoto and M. Nagao. Improvements of japanese morphological analyzer juman. In *Proc. of the Int. Workshop on Sharable Natutal Language Resources*, pages 22–28, 1994.

E. Matusov, N. Ueffing, and H. Ney. Computing consensus translation from multiple machine translation systems using enhanced hypotheses alignment. In *Proc. of the 11th Conf. of the European Chapter of the Association for Computational Linguistics*, pages 33–40, Trento, 2006a.

E. Matusov, R. Zens, D. Vilar, A. Mauser, M. Popovic, S. Hasan, and H. Ney. The RWTH machine translation system. In *TC-STAR Workshop on Speech-to-Speech Translation*, pages 31–36, Barcelona, Spain, June 2006b.

I. McCowan, D. Moore, J. Dines, D. Gatica-Perez, M. Flynn, P. Wellner, and H. Bourlard. On the use of information retrieval measures for speech recognition evaluation. IDIAP-RR 73, IDIAP, Martigny, Switzerland, 2004.

I. Melamed. Automatic evaluation and uniform filter cascades for inducing n-best translation lexicons. In *In Third Workshop on Very Large Corpora (WVLC3)*, Boston, 1995.

A. Menezes and C. Quirk. Microsoft research treelet translation system: IWSLT evaluation. In *Proc.of the Int. Workshop on Spoken Language Translation, IWSLT'05*, 2005.

J.A. Nelder and R. Mead. A simplex method for function minimization. *The Computer Journal*, 7:308–313, 1965.

S. Nießen and H. Ney. Morpho-syntactic analysis for reordering in statistical machine translation. *Proc. of the MT-Summit VII*, pages 247–252, September 2001.

T. Nomoto. Multi-engine machine translation with voted language model. In *Proc. of the 42th Annual Meeting of the Association for Computational Linguistics*, pages 494–501, 2004.

F. J. Och. *Statistical Machine Translation: From Single-Word Models to Alignment Templates.* PhD thesis, RWTH Aachen University, Aachen, Germany, October 2002.

F. J. Och and H. Ney. A comparison of alignment models for statistical machine translation. In *Proc. of the 18th conference on Computational linguistics*, pages 1086–1090, Morristown, NJ, USA, 2000.

F.-J. Och, D. Gildea, S. Khudanpur, A. Sarkar, K. Yamada, A. Fraser, S. Kumar, L. Shen, D. Smith, K. Eng, V. Jain, Z. Jin, and D. Radev. A smorgasbord of features for statistical machine translation. In *Proc. of the Human Language Technology Conf., HLT-NAACL'04*, pages 161–168, 2004.

F.J. Och. An efficient method for determining bilingual word classes. In *Proc. of the 9th Conf. of the European Chapter of the Association for Computational Linguistics*, pages 71–76, June 1999.

F.J. Och. Minimum error rate training in statistical machine translation. In *Proc. of the 41th Annual Meeting of the Association for Computational Linguistics*, pages 160–167, Sapporo, July 2003.

F.J. Och and H. Ney. Discriminative training and maximum entropy models for statistical machine translation. In *Proc. of the 40th Annual Meeting of the Association for Computational Linguistics*, pages 295–302, Philadelphia, USA, July 2002.

F.J. Och and H. Ney. A systematic comparison of various statistical alignment models. *Computational Linguistics*, 29(1):19–51, March 2003.

F.J. Och and H. Ney. The alignment template approach to statistical machine translation. *Computational Linguistics*, 30(4):417–449, December 2004.

K. Papineni, S. Roukos, T. Ward, and W-J. Zhu. Bleu: A method for automatic evaluation of machine translation. In *Proc. of the 40th Annual Meeting of the Association for Computational Linguistics*, pages 311–318, Philadelphia, PA, July 2002.

M. Paul. Overview of the IWSLT 2006 Evaluation Campaign. In *Proc. of the International Workshop on Spoken Language Translation*, pages 1–15, Kyoto, Japan, 2006.

M. Paul, T. Doi, Y. Hwang, K. Imamura, H. Okuma, and E. Sumita. Nobody is perfect: ATR's hybrid approach to spoken language translation. In *Proc.of the Int. Workshop on Spoken Language Translation, IWSLT'05*, 2005.

M. Popovic and H. Ney. Pos-based word reorderings for statistical machine translation. In *5th Int. Conf. on Language Resources and Evaluation, LREC'06*, pages 1278–1283, Genoa, May 2006.

M. Przybocki, G. Sanders, and A. Le. Edit distance: A metric for machine translation evaluation. *5th Int. Conf. on Language Resources and Evaluation, LREC'06*, pages 2038–2043, May 2006.

C. Quirk, A. Menezes, and C. Cherry. Dependency treelet translation: Syntactically informed phrasal SMT. In *Proc. of the 43rd Annual Meeting of the Association for Computational Linguistics (ACL'05)*, pages 271–279, Ann Arbor, Michigan, June 2005.

M. R. Costa-jussà and J. A. R. Fonollosa. Analysis of atatistical and morphological classes to generate weighted reordering hypotheses on a statistical machine translation system. In *Annual Meeting of the Association for Computational Linguistics: Proc. of the Second Workshop on Statistical Machine Translation (WMT)*, pages 171–176, Prague, June 2007.

M. Rajman and T. Hartley. Automatically predicting MT systems rankings compatible with fluency, adequacy or informativeness scores. In *In Proc. of the Workshop on Machine Translation Evaluation: Who Did What To Whom*, pages 29–34, Santiago de Compostela, Spain, 2001.

A.-V.I. Rosti, N.F. Ayan, S B. Xiang, R. Schwartz Matsoukas, and B.J. Dorr. Combining outputs from multiple machine translation systems. In *Proc. of the Human Language Technology Conf., HLT-NAACL'07*, pages 228–235, Rocherster, USA, May 2007.

H. Schwenk. Continuous space language models. *Computer Speech and Language*, 21:492–518, 2007.

H. Schwenk, D. Déchelotte, and J. Gauvain. Continuous space language models for statistical machine translation. In *Proceedings of the COLING/ACL 2006 Main Conference Poster Sessions*, pages 723–730, 2006.

C.E. Shannon. Prediction and entropy of printed english. *Bell Sys. Tech. Journal*, 30:50–64, 1951.

C.E. Shannon and W. Weaver. The mathematical theory of communication. *University of Illinois Press, Urbana, IL*, 1949.

L. Shen, A. Sarkar, and F.J. Och. Discriminative reranking for machine translation. In Daniel Marcu Susan Dumais and Salim Roukos, editors, *Proc. of the Human Language Technology Conf., HLT-NAACL'04*, pages 177–184, Boston, Massachusetts, USA, May 2004.

K. C. Sim, W. J. Byrne, M. J.F. Gales, H. Sahbi, and P. C. Woodland. Consensus network decoding for statistical machine transla- tion system combination. In *Proc. of the ICASSP*, volume 4, pages 105–108, Rocherster, USA, 2007.

M. Snover, B. Dorr, R. Schwartz, J. Makhoul, L. Micciula, and R. Weischedel. A study of translation error rate with targeted human annotation. Technical Report LAMP-TR-126,CS-TR-4755,UMIACS-TR-2005-58, University of Maryland, College Park and BBN Technologies, July 2005.

M. Snover, B. Dorr, R. Schwartz, L. Micciulla, and J. Makhoul. A study of translation edit rate with targeted human annotation. In *Proc. Assoc. for Machine Trans. in the Americas*, 2006.

A. Stolcke. SRILM - an extensible language modeling toolkit. In *Proc. of the 7th Int. Conf. on Spoken Language Processing, ICSLP'02*, pages 901–904, Denver, USA, September 2002.

T. Takezawa, E. Sumita, F. Sugaya, H. Yamamoto, and S. Tamamoto. Toward a broad coverage bilingual corpus for speech translation of travel conversations in the real world. In *3th Int. Conf. on Language Resources and Evaluation, LREC'02*, pages 147–152, Las Palmas, 2002.

C. Tillmann. A unigram orientation model for statistical machine translation. In *Proc. of the Human Language Technology Conf., HLT-NAACL'04*, pages 101–104, Boston, May 2004.

C. Tillmann and H. Ney. Word reordering and a dynamic programming beam search algorithm for statistical machine translation. *Computational Linguistics*, 29(1):97–133, March 2003.

C. Tillmann and T. Zhang. A localized prediction model for statistical machine translation. In *Proc. of the 43th Annual Meeting of the Association for Computational Linguistics*, Michigan,USA, 2005.

H. Tsukada, T. Watanabe, J. Suzuki, H. Kazawa, and H. Isozaki. The NTT statistical machine translation system for IWSLT2005. In *Proc.of the Int. Workshop on Spoken Language Translation, IWSLT'05*, 2005.

E. Vidal. Finite-state speech-to-speech translation. In *Proc. Int. Conf. on Acoustics Speech and Signal Processing*, pages 111–114, Munich, April 1997.

D. Vilar, J. Xu, L. F. D'Haro, and H. Ney. Error Analysis of Machine Translation Output. In *5th Int. Conf. on Language Resources and Evaluation, LREC'06*, pages 697–702, Genoa, Italy, May 2006.

S. Vogel, H. Ney, and C. Tillmann. HMM-based word alignment in statistical translation. In *Proc. of the 16th International Conference on Computational Linguistics*, pages 836–841, Copenhagen, Denmark, August 1996.

C. Wang, M. Collins, and P. Koehn. Chinese syntactic reordering for statistical machine translation. In *Proc. of the 2007 Joint Conf. on Empirical Methods in Natural Language Processing and Computational Natural Language Learning (EMNLP-CoNLL)*, pages 737–745, Prague, June 2007.

T Watanabe, H. Tsukada, and H. Isozaki. Left-to-right target generation for hierarchical phrase-based translation. In *Proc. of the COLING-ACL'06*, pages 777–784, Sydney, Australia, July 2006.

W. Weaver. Translation. In W.N. Locke and A.D. Booth, editors, *Machine Translation of Languages*, pages 15–23. MIT Press, Cambridge, MA, 1955.

B. Wellington, S. Waxmonsky, and I. D. Melamed. Empirical lower bounds on the complexity of translational equivalence. In *Proc. of the 21st International Conference on Computational Linguistics and 44th Annual Meeting of the Association for Computational Linguistics*, pages 977–984, Sydney, Australia, July 2006.

D. Wu. Stochastic inversion transduction grammars and bilingual parsing of parallel corpora. *Computational Linguistics*, 23(3):377–403, September 1997.

D. Wu. A polynomial-time algorithm for statistical machine translation. In *Annual Meeting of the Association for Computational Linguistics*, Santa Cruz, June 1996.

F. Xia and M. McCord. Improving a statistical mt system with automatically learned rewrite patterns. In *Proc. of the 20th International Conference on Computational Linguistics*, page 508, Morristown, 2004.

D. Xiong, Q. Liu, and S. Lin. Maximum entropy based phrase reordering model for statistical machine translation. In *Proc. of the 21st International Conference on Computational Linguistics and 44th Annual Meeting of the Association for Computational Linguistics*, pages 521–528, Sydney, Australia, July 2006.

K. Yamada and K. Knight. A decoder for syntax-based statistical MT. *Proc. of the 40th Annual Meeting of the Association for Computational Linguistics*, pages 303–310, July 2002.

R. Zens and H. Ney. Improvements in phrase-based statistical machine translation. In *Proc. of the Human Language Technology Conf., HLT-NAACL'04*, pages 257–264, 2004.

R. Zens and H. Ney. A comparative study on reordering constraints in statistical machine translation. In *Annual Meeting of the Association for Computational Linguistics*, pages 144–151, Sapporo, July 2003.

R. Zens, F.J. Och, and H. Ney. Phrase-based statistical machine translation. In M. Jarke, J. Koehler, and G. Lakemeyer, editors, *KI - 2002: Advances in artificial intelligence*, volume LNAI 2479, pages 18–32. Springer Verlag, September 2002.

H. Zhang, H. Yu, D. Xiong, and Q. Liu. HMM-based chinese lexical analyzer ICTCLAS. In *Proc. of the 2nd SIGHAN Workshop on Chinese language processing*, pages 184–187, Sapporo, Japan, 2003.

Y. Zhang, R. Zens, and H. Ney. Chunk-level reordering of source language sentences with automatically learned rules for statistical machine translation. In *Proc. of the Human Language Technology Conf. (HLT-NAACL'06):Proc. of the Workshop on Syntax and Structure in Statistical Translation (SSST)*, pages 1–8, Rochester, April 2007.

L. Zhou, C. Lin, and E. Hovy. Re-evaluating machine translation results with paraphrase support. In *Proc. of the Conf. on Empirical Methods in Natural Language Processing, EMNLP'06*, Sydney, Australia, 2006.

www.ingramcontent.com/pod-product-compliance
Lightning Source LLC
LaVergne TN
LVHW042334060326
832902LV00006B/164